Cognitive Fusion for Target Tracking

Synthesis Lectures on Algorithms and Software in Engineering

Editor
Andreas Spanias, *Arizona State University*

Cognitive Fusion for Target Tracking
Ioannis Kyriakides
2019

Secure Sensor Cloud
Vimal Kumar, Amartya Sen, and Sanjay Madria
2018

Sensor Analysis for the Internet of Things
Michael Stanley and Jongmin Lee
2018

Virtual Design of an Audio Lifelogging System: Tools for IoT Systems
Brian Mears and Mohit Shah
2016

Despeckle Filtering for Ultrasound Imaging and Video, Volume II: Selected Applications, Second Edition
Christos P. Loizou and Constantinos S. Pattichis
2015

Despeckle Filtering for Ultrasound Imaging and Video, Volume I: Algorithms and Software, Second Edition
Christos P. Loizou and Constantinos S. Pattichis
2015

Latency and Distortion of Electromagnetic Trackers for Augmented Reality Systems
Henry Himberg and Yuichi Motai
2014

Bandwidth Extension of Speech Using Perceptual Criteria
Visar Berisha, Steven Sandoval, and Julie Liss
2013

Control Grid Motion Estimation for Efficient Application of Optical Flow
Christine M. Zwart and David H. Frakes
2013

Sparse Representations for Radar with MATLAB ™ Examples
Peter Knee
2012

Analysis of the MPEG-1 Layer III (MP3) Algorithm Using MATLAB
Jayaraman J. Thiagarajan and Andreas Spanias
2011

Theory and Applications of Gaussian Quadrature Methods
Narayan Kovvali
2011

Algorithms and Software for Predictive and Perceptual Modeling of Speech
Venkatraman Atti
2011

Adaptive High-Resolution Sensor Waveform Design for Tracking
Ioannis Kyriakides, Darryl Morrell, and Antonia Papandreou-Suppappola
2010

MATLAB™ Software for the Code Excited Linear Prediction Algorithm: The Federal Standard-1016
Karthikeyan N. Ramamurthy and Andreas S. Spanias
2010

OFDM Systems for Wireless Communications
Adarsh B. Narasimhamurthy, Mahesh K. Banavar, and Cihan Tepedelenliouglu
2010

Advances in Modern Blind Signal Separation Algorithms: Theory and Applications
Kostas Kokkinakis and Philipos C. Loizou
2010

Advances in Waveform-Agile Sensing for Tracking
Sandeep Prasad Sira, Antonia Papandreou-Suppappola, and Darryl Morrell
2008

Despeckle Filtering Algorithms and Software for Ultrasound Imaging
Christos P. Loizou and Constantinos S. Pattichis
2008

Cognitive Fusion for Target Tracking

Ioannis Kyriakides

ISBN: 978-3-031-00400-1 paperback
ISBN: 978-3-031-01528-1 ebook
ISBN: 978-3-031-00016-4 hardcover

DOI 10.1007/978-3-031-01528-1

A Publication in the Springer series
SYNTHESIS LECTURES ON ALGORITHMS AND SOFTWARE IN ENGINEERING

Lecture #19
Series Editor: Andreas Spanias, *Arizona State University*
Series ISSN
Print 1938-1727 Electronic 1938-1735

Cognitive Fusion for Target Tracking

Ioannis Kyriakides
University of Nicosia

SYNTHESIS LECTURES ON ALGORITHMS AND SOFTWARE IN ENGINEERING #19

ABSTRACT

The adaptive configuration of nodes in a sensor network has the potential to improve sequential estimation performance by intelligently allocating limited sensor network resources. In addition, the use of heterogeneous sensing nodes provides a diversity of information that also enhances estimation performance. This work reviews cognitive systems and presents a cognitive fusion framework for sequential state estimation using adaptive configuration of heterogeneous sensing nodes and heterogeneous data fusion. This work also provides an application of cognitive fusion to the sequential estimation problem of target tracking using foveal and radar sensors.

KEYWORDS

Bayesian target tracking, particle filtering, sequential Monte Carlo methods, cognitive fusion, sensor networks

Contents

CHAPTER 1

Introduction

Artificial cognition [1] is inspired by human and animal cognition which have evolved to efficiently provide situational awareness. Sequential estimation systems can utilize the benefits of cognition for the efficient use of often scarce resources, such as sensing, processing, communications, actuation, and power. For example, cognition is especially useful when sensing nodes are deployed in areas at land, air, or sea, where resources are scarce, for applications including the estimation of ocean surface currents, and tracking animals, humans, or vehicles. Using a cognitive approach, sensing nodes can be configured to improve the tracking performance of multiple targets, to direct resources toward tracking a specific target of interest in the presence of other targets, and estimate sea currents for ship path planning. Configurable sensing node parameters include, for example, sensor position, sensor orientation, and type of processing to be performed on board of the node. In addition to cognition, the use of heterogeneous nodes in a sensor network (see Figure 1.1) has the potential to enrich information on the evolving state and provide further improvements in estimation performance. Cognition for sequential state estimation consists of a process of configuring the settings of the estimation system based on a prediction of the evolution of the estimated state [2]. Cognition requires the use of practical configurable systems and a framework that utilizes sequentially updated information to derive configuration settings that will improve information acquisition at the next time step of the estimation scenario.

The potential for improvement in estimation performance through the adaptation of settings of configurable systems based on observations on the evolving state is widely acknowledged in the literature. Relevant literature is reviewed in this chapter that provides the general concepts related to cognitive systems and examples of sequential estimation scenarios using cognition. Cognition has been recognized as a driving force for next generation dynamic systems [1, 3–8] in view of technological advancements that enable reconfigurable transmission [9], agile sensing [10], reconfigurable processing [11], adaptive communications schemes [5], or allocation of other essential resources such as power [12, 13]. Cognition has the potential to configure the diverse sensing node system parameters [14–17] to improve information gain based on limited resources.

The Bayesian approach is a suitable framework for cognitive sequential state estimation [1]. The Bayesian method predicts the value of the evolving state being estimated based on existing information and updates the prediction using newly arriving measurements [1]. The prediction of the evolution of the state is then used to configure parameters of nodes in a sensor

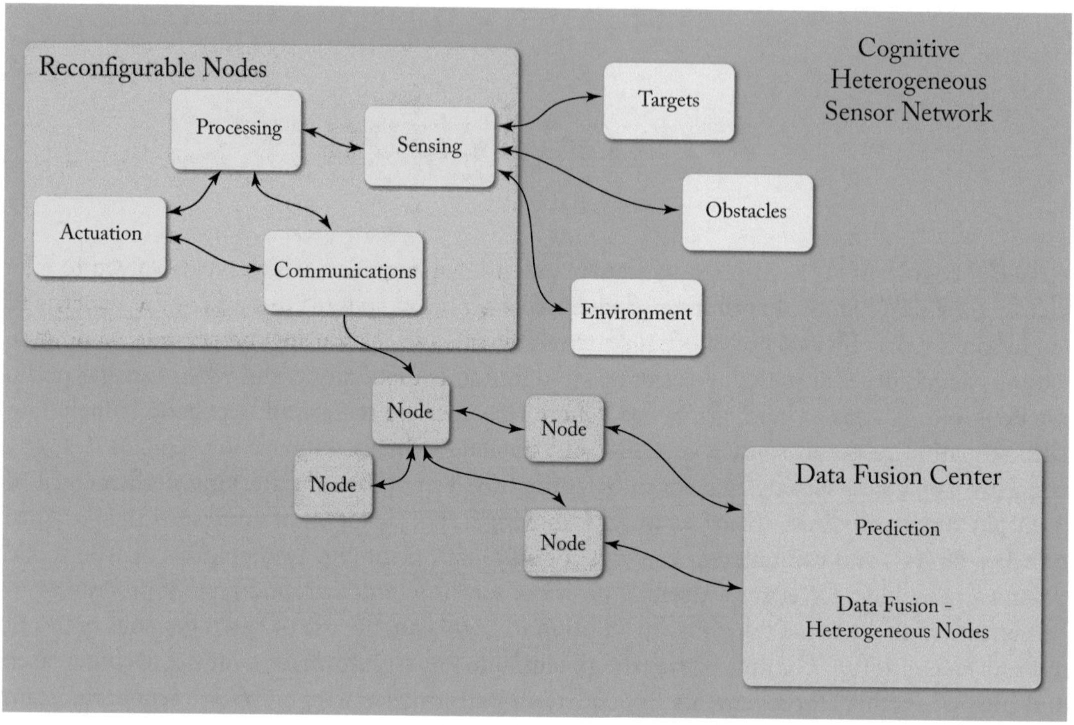

Figure 1.1: A representation of a cognitive heterogeneous sensor network. The fusion center gathers information from multiple sensing nodes and obtains an estimate of the evolving state. The result of estimation is used to predict the future state that is converted to configuration commands distinct for each node in order to improve information acquisition at the next time step.

network to allocate resources toward obtaining more information to improve estimation. The Bayesian approach has been adopted in [18–20] where a cognitive radar framework for target tracking is presented. The method combines estimation of the state of the target with an optimization mechanism to generate node configuration settings based on the predicted target state with the goal of improving tracking performance. The applicability of the general framework is demonstrated via a tracking scenario using a network of sensing nodes where the goal is to improve tracking performance with the individual sensor observation time as a constraint. Target tracking related applications that utilize cognition using the Bayesian methodology include cognitive radar [1, 21, 22] that uses, for example, adaptive waveform design [23–26], while other applications use adaptive sensing [10] and processing [27, 28]. These applications are outlined next.

In an application of radar target tracking with adaptive waveform selection, a prediction of the target state is utilized to adaptively configure the parameters of the transmitted radar waveform [23–25, 29–31]. The goal is to improve tracking performance by configuring parameters to reduce either the expected Mean Squared Error (MSE) [9] or the posterior Cramér-Rao Bound [32, 33]. In addition, Bayesian methods using a sequential Monte Carlo approach [34] have also been applied to the problem of adaptive waveform design [9, 35]. These methods allow the use of accurate measurement models describing a possibly nonlinear relationship to the state. Cognitive radar methods were also applied to achieve dynamic spatial illumination in [36–38] where a beamforming method is used for Multiple Input Multiple Output (MIMO) radar target tracking that focuses the transmit pattern toward the target estimated direction. The method is based on an optimization method using an approximation of the sequential conditional Bayesian Cramér-Rao Bound. Moreover, in [35] adaptive waveform configuration improves tracking performance by positioning radar ambiguity function sidelobes on the radar delay-Doppler plane to unmask weak targets in the presence of strong target returns based on a prediction of the relative target position.

Adaptivity can also be applied at the sensing stage. For example, a foveal sensor [39–42] can be adaptively configured to maintain target presence within the high-acuity region of the sensor for improved target tracking. In [10, 43, 44] the location and acuity of the foveal region is adapted so that higher accuracy measurements are collected in selected directions using the estimated target location and a measure of uncertainty of the estimate. Additionally to sensing, adaptivity in processing can be employed. In [27, 45], a Bayesian method is used to predict the amount of information flow from possible processing actions and select the best action to optimize processor management. In compressive sensing applications, an adaptive compressive sensing radar method is used in [46] where the transmitted waveform and sensing matrix are configured using information available on the target state. Moreover, in [2] the sensing and processing matrix is configured to improve Signal-to-Noise Ratio (SNR) and tracking performance.

To increase information on the estimated state heterogeneous sensing nodes can be used to provide diverse information about the state. However, the ability to configure multiple types of nodes based on an available prediction of the state is a challenging problem, especially when measurements have a nonlinear relationship to the state. A significant challenge in performing accurate data fusion [47] and extraction of information on the target state from heterogeneous sensing nodes is that data from heterogeneous nodes (a) vary in the amount of information they provide and (b) the information they provide is related to diverse characteristics of target, such as identity and kinematic properties. Moreover, nonlinear relationships often exist between measurements and the state [34, 48–51] that result to non-Gaussian posterior probability distributions describing the distribution of the target state given the measurements. Configuring heterogeneous sensing nodes is also challenging due to the diverse configurable characteristics of different nodes. Therefore, cognitive estimation using heterogeneous sensing nodes is a challenging two-fold problem of improving accuracy in state estimation by (a) efficiently fusing data from heterogeneous sensing nodes, and (b) configuring individual nodes based on information from multiple heterogeneous nodes. Although challenging, the solution of this two-fold problem has the potential to improve estimation performance since cognitive heterogeneous nodes both obtain more information on the state and efficiently allocate system resources. Cognitive fusion, a sequential Monte Carlo method is described in this work that is based on the Bayesian estimation framework that enables data fusion and configuration of multiple heterogeneous sensing nodes. Bayesian sequential Monte Carlo methods are reviewed next.

For the sequential estimation problem of tracking targets using measurements that are nonlinearly related to the state and where the posterior distribution of the state is non-Gaussian, particle filtering methods are used that are based on the Bayesian methodology and sequential Monte Carlo methods [34, 48–51]. A basic form of particle filtering, the Sampling Importance Resampling (SIR) particle filter, samples from the state evolution model to form hypotheses on the state which are referred to as particles [34, 48]. However, when estimating high-dimensional states or when using measurements from high-resolution sensors the SIR particle filter becomes highly computationally expensive due to its inability to represent the high dimensionality posterior [49, 50] and the peaked measurement likelihood [48] accurately. For states, arising for example due to the presense of multiple targets in a target tracking application, sequential partition methods [51] have been proposed that form hypotheses concerning individual targets, and perform intermediate weighting and sampling processes from which accurate hypotheses on states of individual targets, called partitions, are selected. Then, accurately generated partitions are combined into particles representing the entire multitarget state. Moreover, sampling from the likelihood distribution [52, 53] is used to effectively generate hypotheses on the state based on information from measurements. The combination of sequential partitions sampling with likelihood sampling has been shown to improve tracking performance compared to the SIR in multitarget and high-resolution applications when using sensing nodes of the same type [54]. However, challenges exist in developing efficient methods of data fusion from heterogeneous sensing nodes having varying resolution and a nonlinear relationship with the target state and the simultaneous configuration of multiple heterogeneous sensors, based on information gathered by a fusion of data from heterogeneous sensors.

Cognitive Fusion, a Bayesian Sequential Monte Carlo method, achieves measurement and node configuration by combining sampling from the likelihood with an auxiliary process of sampling from the state evolution model. Cognitive fusion then draws information from heterogeneous data to configure sensing, processing, communication, actuation, and power related parameters of each sensing node considering their type and individual constraints, with the goal, for example, of reducing ambiguity in the measurements or efficiently process the raw measurements producing minimal size sufficient statistics that can be efficiently communicated to a data fusion center. The method can be considered in combination with random finite sets methods [55–58] that are able to handle a varying number of targets and observations and methods dealing with asynchronous measurements from multiple nodes [59].

This book is organized as follows. Chapter 2 provides the framework of cognitive fusion. Chapter 3 presents the implementation of cognitive fusion for target tracking utilizing a fusion of measurements and configuration of foveal and radar sensing nodes. Chapter 4 provides conclusions. Table 1.1 provides the main notation used in this work.

Table 1.1: Main notation (*Continues.*)

Notation	Description
$\mathbf{x}_k$	State vector at time step k.
$g(\cdot)$	Function that describes the evolution of the state.
$p(\mathbf{x}_k \| \mathbf{x}_{k-1})$	State evolution distribution.
$\mathcal{B}, \beta_u$	Set of possible node parameter settings for all nodes and for node u, respectively.
$\mathbf{B}_k$, $\mathbf{b}_{u,k}$	Selected node parameter settings for all nodes and for node u, respectively after optimization.
$\bar{l}_{u,k}$, $l_{u,k}$	Index corresponding to a resolution cell in the measurement space of node u where true state $\mathbf{x}_k$ and a hypothesis on the target state $\mathbf{x}_{n,k}$ map onto, respectively.
$\mathbf{r}_{\mathbf{b},\bar{l}u,k}$	Measurement vector at node u indexed by $\bar{l} \in \mathcal{L}_u$.
$f_{\mathbf{b},u,k}(\cdot)$	Mapping from state $\mathbf{x}_k$ to measurement $\mathbf{r}_{\mathbf{b},\bar{l}u,k}$ at node u with parameters $\mathbf{b}_{u,k}$.
$f'_{\mathbf{b},u,k}(\cdot)$	Mapping from state $\mathbf{x}_k$ to vector $\mathbf{s}_{\mathbf{b},l,u,k}$ at node u with parameters $\mathbf{b}_{u,k}$.
$\mathbf{s}_{l,u,k}$	Template signal.
$\mathbf{Y}_{\mathbf{b},k}$, $\mathbf{y}_{\mathbf{b},u,k}$, $y_{\mathbf{b},u,k}(\bar{l},l)$, $\bar{l}, l \in \mathcal{L}_u$, $u = 1,\ldots,U$	Multinode, single node, and single-resolution cell measurement statistics.
$\eta_{\bar{l},l,u,k}$	Signal to Noise Ratio (SNR).
$p(\mathbf{Y}_{\mathbf{B},k} \| \mathbf{x}_k)$, $p_u(\mathbf{y}_{\mathbf{b},u,k} \| \mathbf{x}_k)$, $p_u(\breve{y}_{\mathbf{b},u,k}(\bar{l}, l) \| \mathbf{x}_k)$	Likelihood for multiple nodes, single node, and single measurement.
$\breve{p}_u(\mathbf{y}_{\mathbf{b},u,k} \| 0)$	Likelihood in the noise only case.
$\Lambda_{\mathbf{b},\bar{l},l,u,k}$, $\lambda \mathrm{b}_{\mathbf{b},\bar{l},l,u,k}$	Normalized and unnormalized likelihood ratio.
$\mathcal{E}_{\mathbf{B}}$	Expected mean squared error (MSE).
$\epsilon_{\mathbf{b},\bar{l},l,u,k}$	Resolution cell error.
$\bar{\epsilon}_{\mathbf{b},\bar{l},l,u,k}$	Expected resolution cell error.
$\bar{\alpha}_{\mathbf{b},u,k}(\bar{l},l)$	Expected ambiguity function sidelobe level.
$p(\mathbf{x}_k \| \mathbf{Y}_{\mathbf{B},k})$	The posterior distribution of the state.
$\hat{\mathbf{x}}_{\mathbf{B},k}$	Minimum mean squared error (MMSE) estimate of the multitarget state.
$\mathbf{x}_{n,k}$	Particle representing a hypothesis on the state with particle index $n = 1, \ldots, N$.

Table 1.1: (*Continued.*) Main notation

Notation	Description
$\mathcal{L}_{\mathbf{b},u,k}$	Set of resolution cells l of node u at time step k that may be activated due to state with value $\mathbf{x}_k$.
$\mathcal{E}_{\mathbf{B},k}$	Expected MSE.
$\bar{\eta}_{l,l,u,k}$	Expected SNR.
$p(\mathbf{x}_k \mid \mathbf{Y}_{k-1})$	The predicted one time step ahead posterior distribution of the state.
$p_{\mathbf{b},u,k}(l)$	The probability with which state $\mathbf{x}_k$ will map to resolution cell with index $l_{u,k}$.
$\hat{\mathbf{x}}_{\mathbf{B},k}$	MSE estimate of the state.
$q(\mathbf{x}_k \mid \mathbf{x}_{k-1}, \mathbf{Y}_{k-1})$	Importance density.
$\omega_{n,k}$, $n = 1, \ldots, N$	Particle weights.

CHAPTER 2

Cognitive Fusion

This chapter provides the state evolution model and the configurable sensing node measurement model and describes the sequential estimation process of cognitive fusion. The state evolution model describes the changes that the state undergoes through time. The measurement model describes the mapping from the state to the measurement space of each sensing node which includes measurement acquisition and pre-processing of the raw measurements resulting to measurement statistics. The measurement model includes configuration parameters affecting sensing node capabilities including measurement acquisition, processing, communication, and actuation that are selected to improve estimation performance based on the prediction on the evolution of the state. The cognitive fusion method for heterogeneous data fusion and heterogeneous node configuration is presented in a sequential Monte Carlo implementation that is able to handle nonlinear, non-Gaussian estimation problems to facilitate practical applications of the method.

2.1 STATE SPACE FORMULATION

2.1.1 STATE EVOLUTION MODEL

The state at time step k is denoted by the vector $\mathbf{x}_k$ and represents characteristics of, for example, the position, velocity, and acceleration of a vehicle. The evolution of the state is in general given by

$$\mathbf{x}_k = g(\mathbf{x}_{k-1}) + \mathbf{V}_{\mathbf{x}} \boldsymbol{\eta}_{\mathbf{x},k}, \tag{2.1}$$

where $\mathbf{x}_{k-1}$ is the state at time step $k-1$ and $g(\cdot)$ is a state evolution function. Moreover, $\boldsymbol{\eta}_{\mathbf{x},k}$ is the process noise vector that is assumed to include independent zero-mean, Gaussian elements with variance placed in the main diagonal in the covariance matrix $\mathbf{V}_{\mathbf{x}}$. The state evolution distribution is then a Gaussian and is denoted as $p(\mathbf{x}_k|\mathbf{x}_{k-1})$.

Moreover, the predicted posterior of the state at one step ahead based on the state evolution model and measurement statistics $\mathbf{Y}_{k-1}$ at time $k-1$ is provided via the Chapman–Kolmogorov equation [48] as

$$p\left(\mathbf{x}_k|\mathbf{Y}_{k-1}\right) = \int_{\mathbf{x}_{k-1}} p\left(\mathbf{x}_k|\mathbf{x}_{k-1}\right) p\left(\mathbf{x}_{k-1}|\mathbf{Y}_{k-1}\right) d\mathbf{x}_{k-1}. \tag{2.2}$$

The predicted posterior provides information on the future state which is used to configure node parameters for improved information acquisition.

2.1.2 MEASUREMENT MODEL

The measurement model described in this section defines the mapping of the state to the measurement space. It includes configurable parameters that potentially change the size of resolution cells, node position, SNR and the ambiguity function surface. The configurable measurement model is outlined below.

Configuration Parameters

Let $\mathcal{B} = \{\beta_{u,k}\}_{u=1}^{U}$ be a collection of parameter variables that describe configuration settings for all U nodes. Set $\mathcal{B}$ contains subsets $\beta_{u,k}$ that include a collection of parameters specific to individual sensing nodes $u = 1, \ldots, U$. Such parameters values are related to actuation, transmission, sensing, and processing depending on the type of reconfigurable node. For example, for an imager node the set β_u may include imager pan-tilt-zoom settings and for a foveal node equipped with a variable resolution set of cells the configurable parameter is the angle of gaze. Moreover, $\mathbf{B}_k = \{\mathbf{b}_{u,k}\}_{u=1}^{U}$ denote the specific selected parameter values selected for nodes at time step k.

Template Signal Vector

Moreover, state vector $\mathbf{x}_k$ is mapped to vector

$$\mathbf{s}_{l,u,k} = f'_{\mathrm{b},u,k}(\mathbf{x}_k) \tag{2.3}$$

in the measurement space of sensor u through a function that does not include random measurement effects. This mapping identifies template signals $\mathbf{s}_{l,u,k}$, $l \in \mathcal{L}_u$ for processing measurements using matched filtering and l is defined as the index of the template signal. In addition, l also denotes the index of a discrete measurement resolution cell of the node. Index l takes values in set $\mathcal{L}_u$ with cardinality $L_u = |\mathcal{L}_u|$ where $\mathcal{L}_u$ represents the set of indices associated with template signals that may appear in the measurement space of each sensor u due to a state with value $\mathbf{x}_k$ and similarly also represent the set of resolution cells of the sensing node. For example, index l may represent a radar return signal parametrized by a pair of delay and Doppler values and, therefore, l linearly indexes the two-dimensional delay-Doppler plane [2].

Prediction on Node Measurement Space

The information available on the state given in (2.2) is mapped to the measurement space of each node u. The probability with which a state $\mathbf{x}_k$ will map to resolution cell with index $l_{u,k}$ of sensing node u at one time step ahead k, based on prior information available from the estimation process, and under configuration $\mathbf{b}_{u,k}$, is given by

$$p_{\mathrm{b},u,k}(l) = \int_{\mathbf{x}_k : \mathbf{s}_{l,u,k} = f'_{\mathrm{b},u,k}(\mathbf{x}_k)} p(\mathbf{x}_k | \mathbf{Y}_{k-1})\, d\mathbf{x}_k \tag{2.4}$$

for $\mathbf{b}_{u,k} \in \beta$, $l = 1, \ldots, \mathcal{L}_u$. From set $\mathcal{L}_u$ a subset $\mathcal{L}_{\mathrm{b},u,k} \in \mathcal{L}_u$ is defined that represents a subset of the indices of template signals in (2.3) and resolution cells within the measurement space of node u where the target state will map to with non-negligible probability $p_{b,u,k}(l)$. Therefore, knowledge of the probability distribution $p_{b,u,k}(l)$ can be used to reduce the number of processing operations, the measurement statistics that need to be communicated, and to optimize the settings of a configurable node to improve estimation performance as shown in this chapter and Annex A. This probability distribution depends on the node configuration settings if the settings include the possibility of varying resolution. This is true, for example, in the case of the foveal node described in Chapter 3.

Ambiguity Function

The ambiguity function [60] is, moreover, defined as

$$\alpha_{b,u,k}\left(\bar{l}, l\right) = \mathbf{s}^*_{\bar{l},u,k}\mathbf{s}_{l,u,k}, l \in \mathcal{L}_{b,u,k}, \tag{2.5}$$

where $\bar{l}$ is the index of the resolution cell on which the true state maps onto, while the ambiguity function takes values over all measurement resolution cells indices in set $\mathcal{L}_{b,u,k}$. Moreover, $\alpha_{b,u,k}(\bar{l}, l)$ for $l = \bar{l}$ is the mainlobe and $\alpha_{b,u,k}(\bar{l}, l)$ for $l \neq \bar{l} \in \mathcal{L}_{b,u,k}$ are the sidelobes.

Measurement Vector

State $\mathbf{x}_k$ given in (2.1) is mapped to a measurement vector in the measurement space of node u as

$$\mathbf{r}_{b,\bar{l},u,k} = f_{b,u,k}\left(\mathbf{x}_k\right) \tag{2.6}$$

through function $f_{b,u,k}(\cdot)$ that depends on the selected actuation, sensing, and processing settings in set $\mathbf{b}_{u,k}$ for nodes $u = 1, \ldots, U$ and includes random measurement elements. In addition, the measurement vector $\mathbf{r}_{b,\bar{l},u,k}$ is indexed by $\bar{l} \in \mathcal{L}_{b,u,k}$. Moreover, in the noise only case then the measurement vector is given by

$$\breve{\mathbf{r}}_{b,0,u,k} = f_{b,u,k}(0). \tag{2.7}$$

2.1.3 MEASUREMENT PROCESSING

Vectors $\mathbf{s}_{l,u,k}$, $l \in \mathcal{L}_{b,u,k}$ that represent a set of template signals are used for processing measurements $\mathbf{r}_{b,\bar{l},u,k}$ in (2.6) arriving at node u resulting to measurement statistics

$$y_{b,u,k}\left(\bar{l}, l\right) = \mathbf{s}^*_{l,u,k}\mathbf{r}_{b,\bar{l},u,k}, l = 1, \ldots, L_{b,u,k} \tag{2.8}$$

and, in the noise only case, as

$$\breve{y}_{b,u,k}(0, l) = \mathbf{s}^*_{l,u,k}\breve{\mathbf{r}}_{b,0,u,k}, l = 1, \ldots, L_{b,u,k}. \tag{2.9}$$

In addition, $\mathbf{y}_{b,u,k} = [y_{b,u,k}(\bar{l}, l)]_{l=1}^{L_{u,k}}$ represents a vector of measurement statistics for each sensor $u = 1, \ldots, U$ and $\mathbf{Y}_{B,k} = \{\mathbf{y}_{b,u,k}\}_{u=1}^{U}$ represents the multinode measurement statistics with $\breve{\mathbf{y}}_{b,u,k}$ and $\breve{\mathbf{Y}}_{B,k}$ representing the noise only case.

SNR

The signal-to-noise ratio at resolution cell indexed by l is defined by the ratio of the expected values of the signal component to the noise component in the measurements with $\bar{l} = l$ by

$$\eta_{l,l,u,k} = \frac{\sigma_y^2}{\breve{\sigma}_y^2}, l = 1, \ldots, L_{b,u,k}, \tag{2.10}$$

where σ_y^2 and $\breve{\sigma}_y^2$ are the variances of the measurement statistics given in (2.8) and (2.9) when nodes observe the state and in the noise only case, respectively.

Measurement Likelihood

The multinode measurement likelihood is denoted as $p(\mathbf{Y}_{B,k}|\mathbf{x}_k)$. The single sensor likelihood is denoted as $p_u(\mathbf{y}_{b,u,k}|\mathbf{x}_k)$ and the single element likelihood is denoted as $p_u(y_{b,u,k}(\bar{l}, l)|\mathbf{x}_k)$. Based on the assumption that measurements arriving at different sensors are independent the multisensor measurement likelihood is assumed to factorize into the individual likelihoods of each node u as

$$p\left(\mathbf{Y}_{B,k}|\mathbf{x}_k\right) = \prod_{u=1}^{U} p_u\left(\mathbf{y}_{b,u,k}|\mathbf{x}_k\right). \tag{2.11}$$

The statistical independence of the measurements assumption is expected to be accurate since, by design, different nodes in a sensor network are chosen, positioned, and otherwise configured to provide data that provide a diversity of information on the state and, therefore, produce statistically independent measurements. In case of statistical dependence between measurements of different nodes then joint measurement likelihoods can be used for the specific nodes. It is assumed in this work that the state maps onto a single measurement resolution cell indexed by l. Based on the above assumption the measurement likelihood factorizes into single-resolution cell likelihoods as

$$p_u\left(\mathbf{y}_{b,u,k}|\mathbf{x}_k\right) = \prod_{l \in \mathcal{L}_{b,u,k}} p_u\left(y_{b,u,k}\left(\bar{l}, l_u\right)|\mathbf{x}_k\right), \tag{2.12}$$

where the likelihood $p_u(y_{b,u,k}(\bar{l}, l)|\mathbf{x}_k)$ indicates that the true state is $\mathbf{x}_k$ corresponding to index l (2.3). In addition, the likelihood $p_u(\breve{\mathbf{y}}_{b,u,k})$ is defined that denotes the probability of obtaining a measurement statistic $\breve{\mathbf{y}}_{b,u,k}$ in the noise only case and also factorizes as

$$p_u\left(\breve{\mathbf{y}}_{b,u,k}|0\right) = \prod_{l \in \mathcal{L}_{u,k}} p_u\left(\breve{y}_{b,u,k}\left(\bar{l}, l\right)|0\right). \tag{2.13}$$

The likelihood ratio is obtained by dividing in (2.12) with (2.13) and is proportional to the likelihood $p_u(\mathbf{y}_{b,u,k}|\mathbf{x}_k)$ [61–63] is given as

$$p_u\left(\mathbf{y}_{b,u,k}|\mathbf{x}_k\right) \propto \lambda_{b,\bar{l},l,u,k} = \frac{p_u\left(y_{b,u,k}\left(\bar{l}, l\right)|\mathbf{x}_k\right)}{p_u\left(\breve{y}_{b,u,k}\left(\bar{l}, l\right)|0\right)} \tag{2.14}$$

for $l \in \mathcal{L}_{\mathrm{b},u,k}$ and the likelihood ratio is normalized as

$$\Lambda_{\mathrm{b},\bar{l},l,u,k} = \frac{\lambda_{\mathrm{b},\bar{l},l,u,k}}{\sum_l \lambda_{\mathrm{b},\bar{l},l,u,k}} \tag{2.15}$$

or $l \in \mathcal{L}_{\mathrm{b},u,k}$, $\mathbf{b}_{u,k} \in \beta_{u,k}$, $u = 1, \ldots, U$.

Posterior Distribution

The posterior distribution of the state is estimated [34] by

$$p(\mathbf{x}_k | \mathbf{Y}_k) = \frac{p\left(\mathbf{Y}_{\mathrm{B},k} | \mathbf{x}_k\right) p\left(\mathbf{x}_k | \mathbf{Y}_{k-1}\right)}{p\left(\mathbf{Y}_{\mathrm{B},k} | \mathbf{Y}_{k-1}\right)}. \tag{2.16}$$

Minimum Mean Squared Estimate of the State

The minimum mean squared estimate of the state is then given by

$$\hat{\mathbf{x}}_{\mathrm{B},k} = \int \mathbf{x}_k\, p(\mathbf{x}_k | \mathbf{Y}_k)\, d\mathbf{x}_k, \tag{2.17}$$

where both the posterior and estimate depend on the choice of parameters $\mathbf{B}_k$.

Mean Squared Error

The MSE is given by

$$\mathcal{E}_{\mathrm{B}} = E_{\mathrm{B},k}\left[(\bar{\mathbf{x}}_k - \mathbf{x}_k)^T \mathbf{C} (\bar{\mathbf{x}}_k - \mathbf{x}_k)\right], \tag{2.18}$$

where $\mathbf{C}$ normalizes the units of the state vector and $\mathbf{x}_k$ and $\mathbf{x}_k$ take values in a finite set. In addition, the expectation is taken over the unknown true state $\bar{\mathbf{x}}_k$ that is assumed to be distributed as $p(\bar{\mathbf{x}}_k | \mathbf{Y}_{k-1})$ and over the estimated state $\mathbf{x}_k$ distributed as $p(\mathbf{x}_k | \bar{\mathbf{x}}_k \mathbf{Y}_{\mathrm{B},k}, \mathbf{Y}_{k-1})$ where $\mathbf{Y}_{\mathrm{B},k}$ denotes measurements received under configuration $\mathbf{B}_k$ when the true state is $\bar{\mathbf{x}}_k$.

Resolution Cell Error

In Appendix A the resolution cell error is defined for values of the states that map to the same resolution cell of node u for $l = \bar{l}$ as

$$\epsilon_{\mathrm{b},l,l,u,k} = \int_{\bar{\mathbf{x}}_k : \mathbf{s}_{l,u,k} = f'_{\mathrm{b},u,k}(\bar{\mathbf{x}}_k)} \int_{\mathbf{x}_k : \mathbf{s}_{l,u,k} = f'_{\mathrm{b},u,k}(\mathbf{x}_k)} (\bar{\mathbf{x}}_k - \mathbf{x}_k)^T \mathbf{C} (\bar{\mathbf{x}}_k - \mathbf{x}_k)\, d\mathbf{x}_k d\bar{\mathbf{x}}_k, \tag{2.19}$$

which depends on the size of the resolution cell. A sub-optimal method to reduce the overall error is to assign smaller resolution cells to areas of the measurement space on which the state will be mapped to with higher probability. The application of the sub-optimal method depends on the type of cognitive node that possibly allows varying resolution cell sizes, or fixed resolution cells that have different sizes such as the foveal node examined in Chapter 3. Three expectations are defined next based on the definitions of the ambiguity function, SNR, and the resolution error. These expressions provide suboptimal methods to configure sensing nodes to decrease the expected SNR.

Expected Ambiguity Function Sidelobe Level

The expected ambiguity function sidelobe level over resolution cells $l = 1, \ldots, L_{\mathrm{b},u,k}$ given that the true state maps to resolution cell $\bar{l}$ is given by

$$\bar{\alpha}_{\mathrm{b},u,k}\left(\bar{l}, l\right) = \sum_{l \in \mathcal{L}_{\mathrm{b},u,k}, l \neq \bar{l}} \alpha_{\mathrm{b},u,k}\left(\bar{l}, l\right) \; p_{\mathrm{b},u,k}(l) \tag{2.20}$$

for $u = 1, \ldots, U$. The expected ambiguity function sidelobe level expression offers the possibility to shape the ambiguity function surface in order to ensure that low sidelobes exist in the subset of the measurement space of a node where the state will be most likely mapped to. The expected ambiguity function expression can be used to minimize sidelobes over l.

Expected SNR

The expected SNR is given by

$$\bar{\eta}_{l,l,u,k} = \sum_{l \in \mathcal{L}_{\mathrm{b},u,k}} \eta_{l,l,u,k} \; p_{\mathrm{b},u,k}(l) \tag{2.21}$$

for $u = 1, \ldots, U$. Increasing the expected SNR implies the increase of SNR within the subset $\mathcal{L}_{\mathrm{b},u,k}$ of the measurement space of node u where the state is expected to be mapped to at the next time step. The expression can be simplified to increasing the SNR in a subspace of the measurement space associated with high $p_{\mathrm{b},u,k}(l)$. An example of a configurable compressive sensing radar node for which the SNR can be configured to have different values for different resolution cells is provided in Chapter 3.

Expected Resolution Cell Error

The expected resolution cell error is defined as

$$\bar{\epsilon}_{\mathrm{b},l,l,u,k} = \sum_{l \in \mathcal{L}_{\mathrm{b},u,k}} \epsilon_{\mathrm{b},l,l,u,k} \; p_{\mathrm{b},u,k}(l) \tag{2.22}$$

for $u = 1, \ldots, U$. Minimizing the above cost function will, therefore, increase resolution for resolution cells associated with a higher probability $p_{\mathrm{b},u,k}(l)$ (2.4). Therefore, if the node design allows it, the node can be configured to have the state map onto resolution cells with higher resolution with higher probability.

Optimizing the expected values in (2.20), (2.21), and (2.22) is less computationally expensive than minimizing the expected MSE. Further simplifications are possible when the specific characteristics of individual nodes are considered. These simplifications include positioning the node so that the state is in the sensing node's field of view and actively placing the state in high resolution cells of a foveal node as shown in Chapter 3.

2.2 COGNITIVE FUSION FRAMEWORK

This section introduces the cognitive fusion framework that allows heterogeneous data fusion and node configuration. Cognitive fusion combines information from past measurements and the state evolution model to generate a prediction of a range of values that the state may have at one time step ahead with non-negligible probability as shown in the previous section. The predicted range of values of the state then maps onto a set of resolution cells in the measurement space of each of the nodes which depends on the selection of node configuration parameters. The resulting set of resolution cells is a subset of all resolution cells included in the measurement space of each node. Therefore, the prediction on the state produces a prediction on the composition of the measurements that will be collected at each node one time step ahead and is utilized to select node parameters that reduce estimation error. Following the configuration of node parameters and measurement collection, the posterior of the state is estimated. The process described above is represented in a block diagram in Figure 2.1. In realistic scenarios the relationship between the state and the measurements of heterogeneous sensors is diverse and, most often, nonlinear. A sequential Monte Carlo method can then handle nonlinear relationships and non-Gaussian distributions that result from the use of heterogeneous nodes with varying resolution capabilities. This section presents the cognitive fusion framework in a Sequential Monte Carlo implementation. The Sequential Monte Carlo method achieves fusion of heterogeneous node data, processing of measurements that have a nonlinear relationship with the state, and configuration of heterogeneous nodes.

2.2.1 PREDICTION

The predicted one time step ahead posterior distribution of the state, having available past measurement statistics $\mathbf{Y}_{k-1}$ at time $k-1$, is obtained using the Chapman–Kolmogorov equation [48] also defined in (2.2) and denoted as $p(\mathbf{x}_k|\mathbf{Y}_{k-1})$. The predicted one time step ahead posterior represents available information on the one time step ahead state that is extracted from the sequential estimation process and is utilized in node configuration. The prediction process is described next.

Let $\mathbf{x}_{n,k}, n = 1, \ldots, N$ represent N hypotheses (particles) [48] on the value of the state at time step k. The posterior predicted given past measurements in (2.2) is evaluated at these N realizations of the state vector. The numerical approximation of values of the predicted posterior given past measurements is then given by

$$p\left(\mathbf{x}_{n,k}|\mathbf{Y}_{k-1}\right) = p\left(\mathbf{x}_{n,k}|\mathbf{x}_{n,k-1}\right) w_{n,k-1} \tag{2.23}$$

for $n = 1, \ldots, N$ where $w_{n,k-1} = p(\mathbf{x}_{n,k-1}|\mathbf{Y}_{k-1}),\ n = 1, \ldots, N$ is the posterior evaluated at values of the state at the previous time step $k-1$ denoted as $\mathbf{x}_{n,k-1},\ n = 1, \ldots, N$. The number of realizations N required to adequately represent the posterior increases with an increase of the variance and dimensionality of the state space and is, thus, chosen based the application. More-

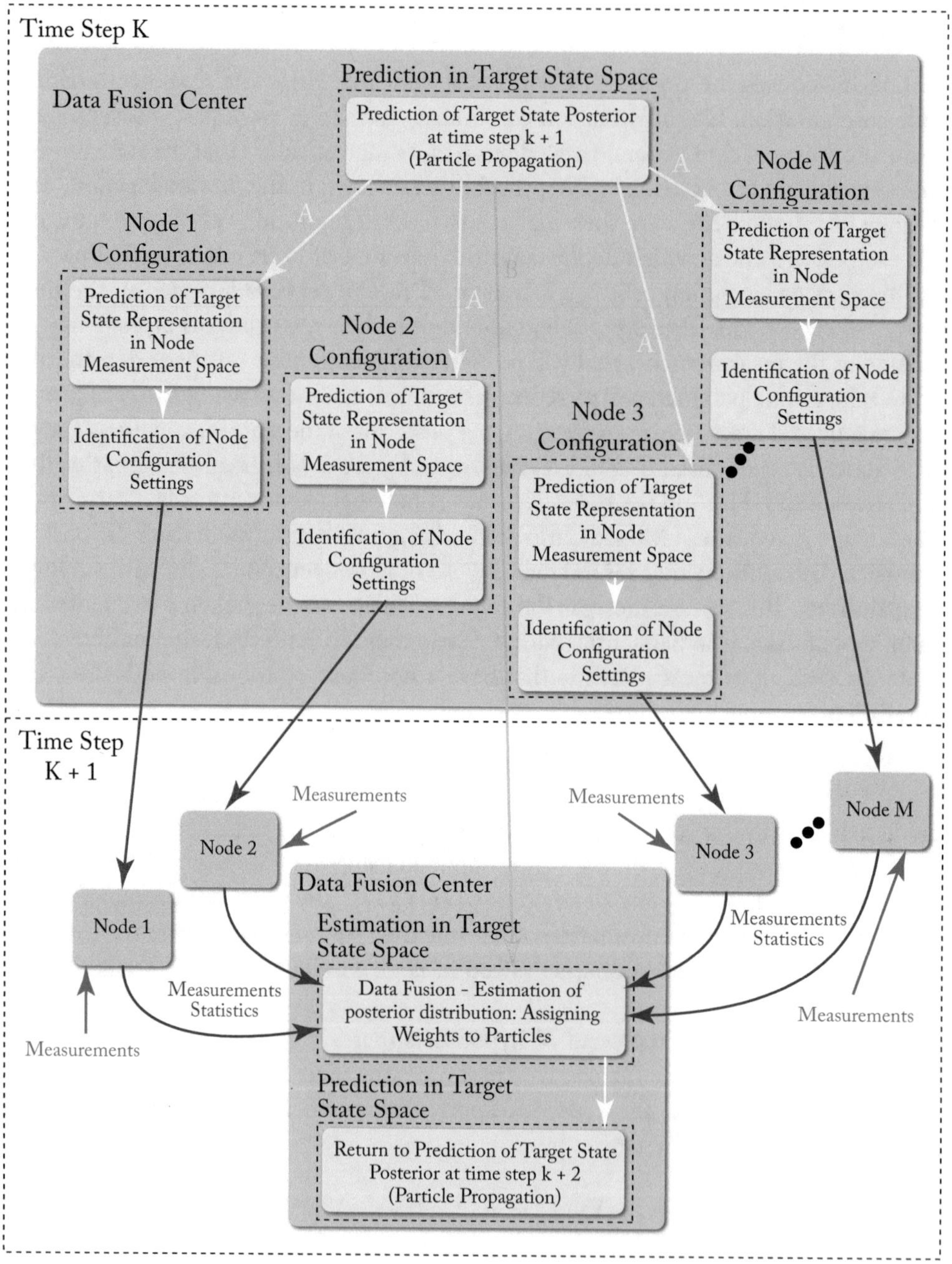

Figure 2.1: Overall process of cognitive fusion using multiple heterogeneous sensor nodes: (a) mapping of prediction from state space to measurement space and (b) production of hypotheses on the state.

over, $\mathbf{Y}_{k-1}$ is a vector of measurement statistics resulting from processing the measurements as described in Section 2.1.3.

Possible future states are next predicted using information from the state evolution model. Each particle $n = 1, \dots, N$ is generated as

$$\mathbf{x}_{n,k} = g\left(\mathbf{x}_{n,k-1}\right) + \mathbf{V}_{\mathrm{x}} \boldsymbol{\eta}_{\mathrm{x},n,k}, \tag{2.24}$$

where vectors $\boldsymbol{\eta}_{\mathrm{x},n,k}, n = 1, \dots, N$ are sampled from a zero-mean unit variance Gaussian distribution and with $g(\cdot)$ and $\mathbf{V}_{\mathrm{x}}$ as defined for (2.1). Equation (2.24) is equivalent to sampling from the state evolution distribution $p(\mathbf{x}_k|\mathbf{x}_{n,k-1})$. The samples $\mathbf{x}_{n,k}, n = 1, \dots, N$, which then represent hypotheses on the state one time step ahead, are mapped to the measurement space of each node under different possible configurations of node parameters. The predictions on state are mapped to the measurement space of each node $u = 1, \dots, U$ under each configuration $\mathbf{b}_{u,k} \in \beta_u$ using (2.3) as

$$\mathbf{s}_{l,u,k} = f'_{\mathrm{b},u,k}\left(\mathbf{x}_{n,k}\right) \tag{2.25}$$

with $n = 1, \dots, N, \mathbf{b}_{u,k} \in \beta_u, u = 1, \dots, U$ where vector $\mathbf{s}_{l,n,u,k}$ represents a template signal indexed by $l_{n,u}$. $f'_{\mathrm{b},u,k}(\cdot)$ denotes a function mapping the state to a signal in the signal space of node u. The set of indices of vectors $s_{l,u,k}$ that may appear at the measurements of each sensor u at time step k and have been obtained under configuration $\mathbf{b}_{u,k}$ and similarly the set of indices of resolution cells on which the true state is likely to mapped onto is identified as

$$\mathcal{L}'_{\mathrm{b},u,k} = \bigcup \{l_{n,u}\}_{n=1}^{N}. \tag{2.26}$$

For high-resolution sensing nodes sampling a limited number of particles N from the kinematic prior may result to discontinuities in the measurement space. As a result the index of the elementary signal in the measurement space of a high-resolution node u that corresponds to the true state may not be included in set $\mathcal{L}'_{\mathrm{b},u,k}$. Due to an insufficient number of samples, set $\mathcal{L}'_{\mathrm{b},u,k}$ will likely be disconnected. A connected set $\mathcal{L}_{\mathrm{b},u,k}$ is generated using interpolation from the possibly disconnected set $\mathcal{L}'_{\mathrm{b},u,k}$ to include the true state related index in the measurements. The information available on the state is next mapped to the measurement space of each node u. The probability with which a state $\mathbf{x}_k$ will map to resolution cell with index $l_{u,k}$ of sensing node u at one time step ahead k, based on prior information available from the estimation process, and under configuration $\mathbf{b}_{u,k}$, is defined in (2.4) as $p_{\mathrm{b},u,k}(l)$ for $\mathbf{b}_{u,k} \in \beta, l = 1, \dots, \mathcal{L}_{\mathrm{b},u,k}$. This probability depends on the node configuration settings if the settings include the possibility of varying resolution. This is true, for example, in the case of the foveal node described in Chapter 3. This probability distribution using a Monte Carlo method under configuration $\mathbf{b}_{u,k}$ is given by

$$p_{\mathrm{b},u,k}(l) = \sum_{n:\mathbf{s}_{l,u,k} = f'_{\mathrm{b},u,k}(\mathbf{x}_{n,k})} w_{n,k-1} \tag{2.27}$$

for $\mathbf{b}_{u,k} \in \beta_{u,k}, l = 1, \dots, \mathcal{L}_{\mathrm{b},u,k}$.

2.2.2 NODE CONFIGURATION

Since the posterior, estimate, and sequential estimation error depend on the choice of parameters $\mathbf{B}_k$ then node parameters can be selected prior to measurement acquisition to minimize the expected MSE [25]. However, the exhaustive process for calculating the expected MSE under each multi-sensor configuration setting $\mathbf{B} \in \mathcal{B}$ with $\mathbf{b}_{u,k} \in \beta_{u,k}, u = 1, \ldots, U$ or performing a computationally intensive optimization method to identify a minimum over a multidimensional space would become prohibitive as the number of nodes grows. A suboptimal solution can be found that configures parameters of each node independently to reduce the expected mean squared error by increasing the expected value of the SNR and reducing the expected value of the ambiguity function and the resolution error. Having the information on the mapping of the state onto the measurement space of each node captured in (2.4) the nodes can be configured by reducing the expected resolution error in (2.22), reducing expected ambiguity function sidelobe levels (2.20), and increasing the expected SNR (2.21). These suboptimal goals are achieved inexpensively by considering the individual characteristics of each sensing node. For example, for a foveal node described in Chapter 3 the configuration process reduces to adjusting the angle of gaze of the node to observe the predicted estimate of the state. In addition, the ambiguity function sidelobes depend on configuration settings of a radar transmitted waveform and are used together with (2.27) to calculate the expected sidelobe level given by (2.20) and select the configuration setting with the least expected ambiguity function sidelobe level. An example of improving expected SNR through the adaptive configuration parameter process is provided in Chapter 3 for adaptive compressive radar. Once parameters are configured for all nodes the measurement collection and update of the posterior of the state are performed as described in the next section.

2.2.3 MEASUREMENT UPDATE

Based on the selected set of parameters $\mathbf{B}_k$ measurements are acquired and processed resulting to measurement statistics $\mathbf{Y}_{\mathbf{B},k}$ as described in Section 2.1.2. Having the resulting measurement statistics available the measurement likelihood $p(\mathbf{Y}_{\mathbf{B},k}|\mathbf{x}_k)$ is constructed. The posterior distribution and the minimum mean squared estimate of the state are then estimated [34] by $p(\mathbf{x}_k|\mathbf{Y}_{\mathbf{B},k})$ in (2.16) and by $\hat{\mathbf{x}}_{\mathbf{B},k}$ in (2.17) where the posterior and estimate depend on the selected parameters $\mathbf{B}_k$. Due to the possibly nonlinear relationship between state and measurements of heterogeneous nodes an analytical evaluation of the expressions involved in the sequential estimation process may not be possible. A Monte Carlo method is then employed to sequentially estimate the non-Gaussian state posterior that results from the nonlinear relationship of the state and heterogeneous nodes as explained next.

The estimate of the posterior distribution of the state depends on the multinode likelihood and the kinematic distribution as shown in (2.16). Therefore, the design of an efficient method for proposing hypotheses on the state must accurately represent both the multinode likelihood and the prior. A sampling method is described in this section that is able to propose samples

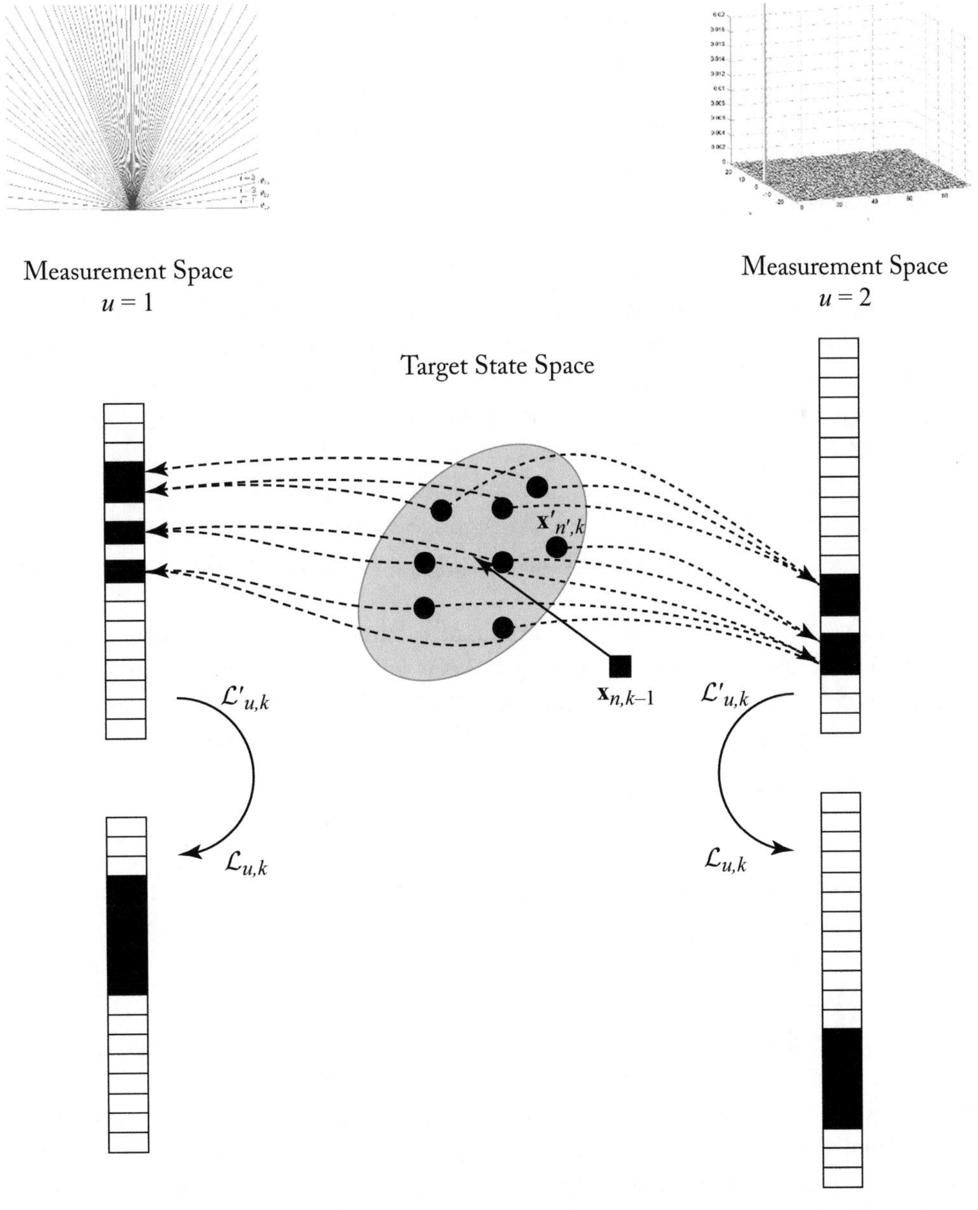

Figure 2.2: Each particle $\mathbf{x}_{n,k-1}$ is projected one time step ahead to produce N' auxiliary particles $\acute{\mathbf{x}}_{n',k}, n' = 1, \ldots, N'$ (2.29). Each of the auxiliary particles are propagated to the measurement space of each sensor (2.25) forming set of indices of signal elements $\mathcal{L}'_{u,k}$ (2.26) for each node which is interpolated to a continuous set $\mathcal{L}_{u,k}$.

of the posterior distribution from both the likelihood associated with multiple nodes and the prior. The sampling process is made efficient via a factorization of the likelihood distribution. The assumption on the factorization of the likelihood distribution does not need to hold with perfect accuracy for the purposes of the sampling process as the bias in proposing particles is taken into account to correct the particle weights as explained later in this section. Next, the two-part sampling method is described: (a) sampling using measurement information and (b) sampling using prior information.

Sampling Using Measurement Information

Sampling using measurement information identifies resolution cells in the measurement space of each node on which the state is more likely to map onto. Sampling from measurement information is especially useful when using high-resolution and heterogeneous nodes. This sampling process is described next. Following acquisition of measurements the resulting measurement vector is given by $\mathbf{r}_{\mathrm{b},\bar{l},u,k} = f_{\mathrm{b},u,k}(\mathbf{x}_k)$ as in (2.6) indexed by $\bar{l}_{u,k}$ where $\mathbf{x}_k$ is the true state at time step k. The measurement vector is processed using templates $\mathbf{s}_{l,u,k}$ as in (2.3) where set of indices $\mathcal{L}_{\mathrm{b},u,k}$ were calculated in (2.26). Processing yields measurement statistics $y_{\mathrm{b},u,k}(\bar{l}, l)$ as in (2.8) and the normalized likelihood ratio $\Lambda_{\mathrm{b},\bar{l},l,u,k}$ as in (2.15) using (2.12), (2.13), and (2.14).

Next, follows a process of sampling indices of vectors in the measurement space of each sensor u from the likelihood ratio where the process is performed independently for each node u allowing parallel implementation. In the process of sampling from the likelihood, an index is sampled from the likelihood ratio for each particle n and for each node u

$$\tilde{l}_{n,u} \sim \left\{\Lambda_{\mathrm{b},\bar{l},l,u,k}\right\}_{l\in\mathcal{L}_{\mathrm{b},u,k}} \tag{2.28}$$

with sampling bias $\tilde{\Lambda}_{\mathrm{b},n,u,k} = \Lambda_{\mathrm{b},\bar{l},l,u,k}|_{l=\tilde{l}_{n,u}}$, i.e., the sampling bias is the likelihood ratio $\Lambda_{\mathrm{b},\bar{l},l,u,k}$ evaluated for $l = \tilde{l}_{n,u}$.

Sampling Using Prior Information

The process of sampling from the measurement space is useful in revealing possible resolution cells on which the state maps onto in the measurement space of individual nodes. However, the mapping from the measurement space to the state space is not, in general, a one-to-one relationship especially in the case of using measurements from multiple heterogeneous nodes. An auxiliary sampling process is then used which selects samples in the state space that correspond to the likelihood sampled measurement space signal indices in (2.28). Specifically, auxiliary samples $n' = 1, \ldots, N'$ are sampled for each particle $n = 1, \ldots, N$ using the state evolution model in (2.1) as

$$\mathbf{x}_{n',k} = g(\mathbf{x}_{n,k-1}) + \mathbf{V}_{\mathrm{x}} \boldsymbol{\eta}_{\mathrm{x},n',k} \tag{2.29}$$

for $n' = 1, \ldots, N'$.

The sampled states are then mapped to the measurement space of each sensor $u = 1, \ldots, U$ to yield signal element $\mathbf{s}_{l,n',u,k}$ in (2.8) indexed by $l_{n,u}$. A sample n'' is then selected out of $n' = 1, \ldots, N'$ such that $n'' = \underset{n'}{\operatorname{argmin}} \sum_{u=1}^{U} |l_{n',u} - \tilde{l}_{n,u}|$ where $\tilde{l}_{n,u}$ is sampled in (2.28). The auxiliary sampling step yields states that correspond to sampled indices of all nodes. Ideally, it would be preferred to have $\sum_{u=1}^{U} |l_{n',u} - \tilde{l}_{n,u}|$ to equal zero for $u = 1, \ldots, U$. However, the auxiliary process may fail to obtain a hypothesis on the state that exactly corresponds to sampled indices of all nodes due to unreliable measurements produced by one node or an insufficient number of particles. To compensate for this event the minimum index discrepancy is selected. In addition, particles that do not correspond to sampled measurement indices will be penalized at the particle weighting step.

When auxiliary sampling is completed and selection of the auxiliary sample index n'' is selected, then $\mathbf{x}_{n,k} = \tilde{\mathbf{x}}_{n'',k}$ is set as a proposed state that is consistent with sampled indices in (2.28). Overall, auxiliary samples are sampled with an importance density

$$q\left(\mathbf{x}_{n,k} | \mathbf{x}_{k-1}, \mathbf{Y}_{k-1}\right) = p\left(\mathbf{x}_{n,k} | \mathbf{x}_{n,k-1}\right) \prod_{u=1}^{U} \tilde{\Lambda}_{\mathrm{b},n,u,k} \tag{2.30}$$

that represents the sampling bias due to sampling from the prior and measurement likelihood.

Particle Weights and Estimation

The weight of each particle [48] is next calculated as

$$w_{n,k} \propto w_{n,k-1} \frac{p\left(\mathbf{x}_{n,k} | \mathbf{x}_{n,k-1}\right)}{q\left(\mathbf{x}_{n,k} | \mathbf{x}_{k-1}, \mathbf{Y}_{k-1}\right)} \prod_{u=1}^{U} \tilde{\Lambda}_{\mathrm{b},n,u,k} \tag{2.31}$$

using the likelihood ratio $\tilde{\Lambda}_{\mathrm{b},n,u,k}$ that was already calculated and used in (2.15), $q(\mathbf{x}_{n,k} | \mathbf{x}_{k-1}, \mathbf{Y}_{k-1})$ is given by (2.30), and the calculated values corresponding to particle $\mathbf{x}_{n,k}$ are calculated given that $p(\mathbf{x}_{n,k} | \mathbf{x}_{n,k-1})$ is Gaussian as described in Section 2.1.1. Using (2.28) and in case $l_{n,u} = \tilde{l}_{n,u}$ for all sensors u the weight expression simplifies to

$$w_{n,k} \propto w_{n,k-1} \tag{2.32}$$

for $n = 1, \ldots, N$ taking into account that each particle $\mathbf{x}_{n,k}$ was sampled in (2.28) with bias $\tilde{\Lambda}_{\mathrm{b},n,u,k}$ and in (2.29) with bias $p(\mathbf{x}_{n,k} | \mathbf{x}_{n,k-1})$. The estimated posterior is given as

$$p\left(\hat{\mathbf{x}}_{\mathbf{B},k} | \mathbf{Y}_{\mathbf{B},k}\right) = \sum_{n=1}^{N} w_{n,k} \delta\left(\mathbf{x}_k - \mathbf{x}_{n,k}\right) \tag{2.33}$$

and the MMSE estimate of the state defined in (2.17) is given by

$$\hat{\mathbf{x}}_{\mathrm{B},k} = \sum_{n=1}^{N} w_{n,k}\mathbf{x}_{n,k} \tag{2.34}$$

expressed in N discrete values of the state. The algorithm is summarized in Table 2.1. After estimation, particles can be resampled to avoid degeneracy, which results when having only few remaining particles with high weights [48].

Table 2.1: Cognitive Fusion

For each time step $k = 1, \ldots, K$

At time step k - 1: Prediction of the future state

- For each particle $n = 1, \ldots, N$
  - o Sample $\mathbf{x}_{n,k} = g(\mathbf{x}_{n,k\text{-}1}) + \mathbf{V}_{\mathbf{x}}\eta_{\mathbf{x},n,k}$ (2.29)
  - o For each node $u = 1, \ldots, U$
    - * For $\mathbf{b}_{u,k} \in \beta_{u,k}$
      - \- Map $\mathbf{s}_{l,u,k} = f'_{\mathbf{b},u,k}(\mathbf{x}_{n,k})$ (2.25)
- For each node $u = 1, \ldots, U$
  - o For $\mathbf{b}_{u,k} \in \beta_{u,k}$
    - * Identify set $\mathcal{L}'_{\mathbf{b},u,k} = \bigcup \{l_{n,u}\}_{n=1}^{N}$ (2.26) and interpolated set $\mathcal{L}_{\mathbf{b},u,k}$
    - * For $l = 1, \ldots, L_{\mathbf{b},u,k}$
      - \- Identify probability $p_{\mathbf{b},u,k}(l) = \sum_{n:\mathbf{s}_{l,u,k} = f'_{\mathbf{b},u,k}(\mathbf{x}_{n,k})} w_{n,k\text{-}1}$ (2.27)

Perform individual node configuration resulting to configuration set $\mathbf{B}_k = \{\mathbf{b}_{u,k}\}_{u=1}^{U}$

At time step k: Measurement Update

- For each node $u = 1, \ldots, U$ using configuration $\mathbf{b}_{u,k}$
  - * Receive measurements $\mathbf{r}_{\mathbf{b},\bar{l},u,k} = f_{\mathbf{b},u,k}(\mathbf{x}_k)$ (2.6)
  - o For $l = 1, \ldots, L_{\mathbf{b},u,k}$
    - * Generate statistics $y_{\mathbf{b},u,k}(\bar{l}, l) = \mathbf{s}^{*}_{l,u,k}\ \mathbf{r}_{\mathbf{b},\bar{l},u,k}$ (2.8)
    - * Calculate likelihood ratio $\Lambda_{\mathbf{b},\bar{l},l,u,k}$ (2.15)
  - o For each particle $n = 1, \ldots, N$
    - * Sample index $\bar{l}_{n,u} \sim \{\Lambda_{\mathbf{b},\bar{l},l,u,k}\}_{l \in \mathcal{L}_{\mathbf{b},u,k}}$ (2.28) with bias $\tilde{\Lambda}_{\mathbf{b},n,u,k} = \Lambda_{\mathbf{b},\bar{l},l,u,k}|_{l=\bar{l}_{n,u}}$
- For each particle $n = 1, \ldots, N$
  - o For each auxiliary particle $n' = 1, \ldots, N$
    - * Sample auxiliary particles $\acute{\mathbf{x}}_{n',k} = g(\mathbf{x}_{n,k\text{-}1}) + \mathbf{V}_{\mathbf{x}}\eta_{\mathbf{x}n',k}$, $n' = 1, \ldots, N$ (2.29)
    - * For each node $u = 1, \ldots, U$
      - \- Map $\mathbf{s}_{l,n',u,k} = f_{\mathbf{b}u}(\mathbf{x}'_{n',k})$, (2.8)
  - o Select $n'' = \underset{n'}{\operatorname{argmin}} \sum_{u=1}^{U} |l_{n',u} - \bar{l}_{n,u}|$, let $\mathbf{x}_{n,k} = \acute{\mathbf{x}}_{n'',k}$,
  - o Calculate weight $w_{n,k} \propto w_{n,k\text{-}1} \dfrac{p(\mathbf{x}_{n,k}|\mathbf{x}_{n,k\text{-}1})}{q(\mathbf{x}_{n,k}|\mathbf{x}_{n,k\text{-}1}\mathbf{Y}_{k\text{-}1})} \prod_{u=1}^{U} \tilde{\Lambda}_{\mathbf{b},n,u,k}$, $n = 1, \ldots, N$ (2.31) and normalize
- Estimate posterior distribution $p(\hat{\mathbf{x}}_{\mathbf{B},k}|Y_{\mathbf{B},k}) = \sum_{n=1}^{N} w_{n,k}\,\delta(\mathbf{x}_k - \mathbf{x}_{n,k})$ (2.33)
- Estimate MMSE state estimate $\hat{\mathbf{x}}_{\mathbf{B},k} = \sum_{n=1}^{N} w_{n,k}\mathbf{x}_{n,k}$ (2.34)

CHAPTER 3

Cognitive Fusion for Target Tracking with Foveal and Radar Nodes

A sequential Monte Carlo method is described in this chapter for tracking a single target using cognitive foveal and radar sensors [64]. The tracking algorithm is based on the cognitive fusion framework described in Chapter 2. The method uses a one time step ahead prediction of the target state to predict the structure of the measurements expected to arrive in each node. This prediction is used to configure foveal and radar nodes to improve target tracking performance.

3.1 STATE SPACE FORMULATION

3.1.1 TARGET MOTION MODEL

A single-point target moves in the two-dimensional Cartesian plane with state described by vector $\mathbf{x}_k = [\chi_k \; \dot{\chi}_k \; \psi_k \; \dot{\psi}_k]^T$ at time step k where χ_k and ψ_k denote the position and $\dot{\chi}_k$ and $\dot{\psi}_k$ denote the velocity of the target. The state evolves based on a constant velocity model as:

$$\mathbf{x}_k = \mathbf{F}\mathbf{x}_{k-1} + \mathbf{V}\boldsymbol{\eta}_k, \tag{3.1}$$

where $\mathbf{F} = [1 \; \delta t \; 0 \; 0; 0 \; 1 \; 0 \; 0; 0 \; 0 \; 1 \; \delta t; 0 \; 0 \; 0 \; 1]$ and δt is the time difference between state transitions. Matrix $\mathbf{V}$ is a diagonal matrix with diagonal elements the square root of the variance of the process noise vector $\boldsymbol{\eta}_k$. The process noise $\boldsymbol{\eta}_k$ is a vector of zero-mean, unit variance Gaussian random variables. Based on this model, the distribution of the state of the target given the state at the previous time step is denoted as $p(\mathbf{x}_k|\mathbf{x}_{k-1})$ and is assumed to be a Gaussian. The foveal and radar sensing node measurement models are presented below.

3.1.2 FOVEAL MEASUREMENT MODEL

The foveal [10, 41–44] node described in this work collects angular measurements with varying resolution. The sensing node is equipped with finer angular resolution cells in the central area of the sensor (foveal region) and more coarse angular resolution cells in the periphery with similar setup to models described in [39, 65, 66] and as illustrated in Figure 3.1. The node is configured by aligning the foveal region toward directions where a target is expected to be found at one time

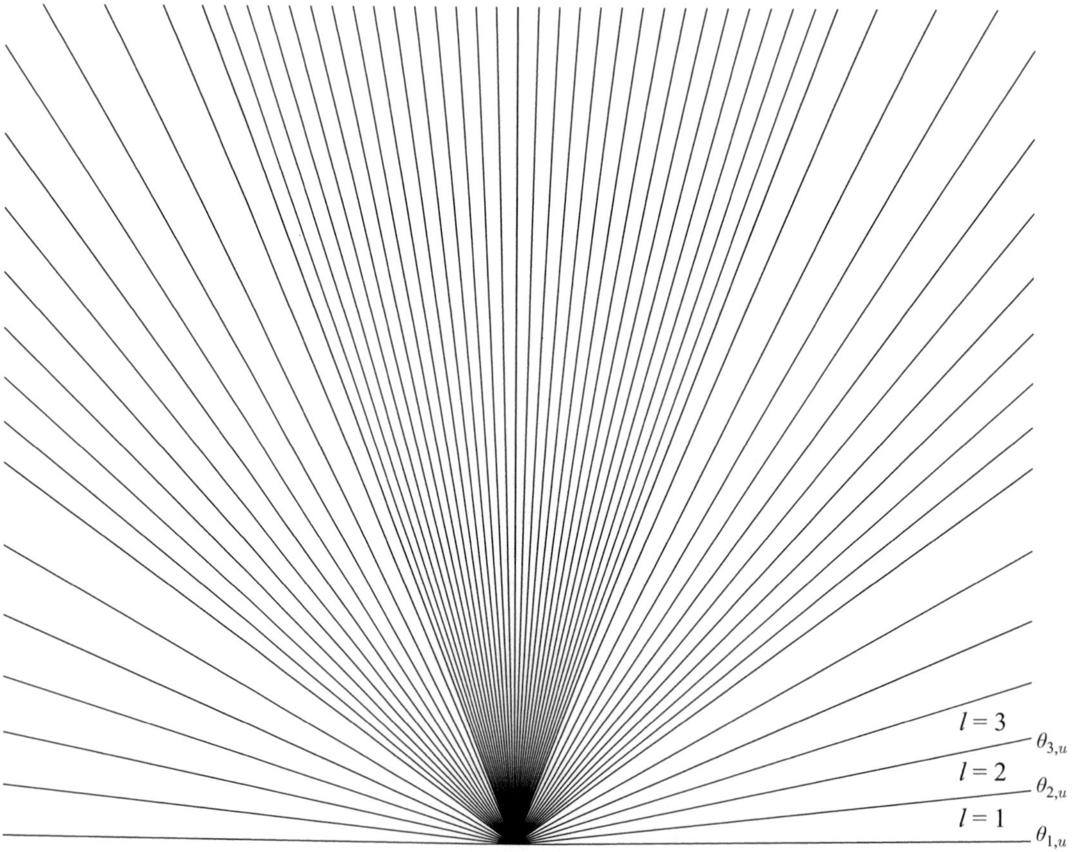

Figure 3.1: Foveal sensor with higher angular resolution in the central area and lower resolution in the periphery.

step ahead. The biologically inspired foveal sensor model [67] offers the benefits of (a) observing a wide area of the scene with low resolution for low data acquisition and processing enabling detection of new targets while preserving resources and (b) observing angular areas where targets are expected to be found with high resolution to obtain more information on specific targets of interest. Therefore, the sensor is efficient in that it allocates less sensing, processing, and communications resources to spatial areas where targets appear more sparsely, while allocating more resources to areas where targets are expected to be found or are persistently observed [68, 69]. This adaptive allocation of resources improves overall performance under resource constraints. The sensor model and measurement acquisition and processing methods are described next.

Each foveal node is separated in angular resolution cells indexed by $l = 1, \ldots, L_u$ where L_u is the total number of resolution cells. The central resolution cell is indexed by $\lfloor \frac{L_u}{2} \rfloor$ and it is configured to point to the angle of gaze $\theta_{o,u,k}$ which is a configuration parameter selected at each time step. Each node covers an angular range of π rad and each cell is oriented at angle $\theta_{l,u} = \theta_{o,u,k} + \sum_{l'=1}^{l}(l'-1)\Delta\theta_l - \frac{\pi}{2}, l \in \mathcal{L}_u$ where $\Delta\theta_l$ is the angular width of resolution cell l. Non-cognitive angular nodes have fixed angular width $\Delta\theta_l = \frac{\pi}{L_u}\ \forall\ l$. The target state is mapped to the measurement space of the foveal sensor by first identifying the angle formed by the target and the sensor as

$$\vartheta_{u,k} = \arctan\frac{\psi_k - y_u}{\chi_k - x_u} - \theta_{o,u,k}, \tag{3.2}$$

where χ_k and ψ_k denote the position of the target at time step k defined in Section 3.1.1 and x_u and y_u denote the position of the foveal sensor. The angle $\vartheta_{u,k}$ is indexed by $l_{u,k} = \underset{l}{\operatorname{argmin}}|\vartheta_{u,k} - \theta_{l,u}|$ by assigning the value of its angle of direction to the nearest resolution cell.

The above mapping is next represented in vector form where an $L_{\mathrm{b},u,k}$-dimensional vector represents template signals and deterministic versions of the measurements is given by

$$\mathbf{s}_{l,u,k} = f'_{\mathrm{b},u,k}(\mathbf{x}_k), l \in \mathcal{L}_u, \tag{3.3}$$

where function $f'_{\mathrm{b},u,k}(\cdot)$ transforms the position of the target described by state $\mathbf{x}_k$ to a vector $\mathbf{s}_{\bar{l},u,k}$ in (3.3) which has value equal to 1 at index l and zero otherwise and depends on configuration parameter $\theta_{o,u,k}$ that represents the angle of gaze. Therefore, the set of configuration settings of the foveal sensor are set as $\mathbf{b}_{u,k} = \{\theta_{o,u,k}, \mathcal{L}_{\mathrm{b},u,k}\}$ where $\mathcal{L}_{\mathrm{b},u,k} \in \mathcal{L}_u$ is a reduced sized set of cells within which the target state is more likely to be mapped to. The $L_{\mathrm{b},u,k}$-dimensional measurement vector has elements

$$\mathbf{r}_{\bar{l},u,k} = f_{\mathrm{b},u,k}(\mathbf{x}_k) = \gamma_{u,k}\mathbf{s}_{\bar{l},u,k} + \boldsymbol{\upsilon}_{l,u,k}, \tag{3.4}$$

where $\gamma_{u,k}$ is a zero-mean complex Gaussian amplitude with variance σ_γ^2 according to the Swerling *I* model [70] and $\boldsymbol{\upsilon}_{l,u,k}, l = 1, \ldots, L_u$ denotes scalar additive zero-mean complex Gaussian noise vector with variance σ_υ^2 and, similarly to the definition in (3.3), $\mathbf{s}_{\bar{l},u,k}$ is a vector with value of 1 at index $\bar{l} = l_{u,k}$.

3.1.3 COMPRESSIVE COGNITIVE RADAR MEASUREMENT MODEL

Active radar sensors collect information on the range and range rate of the target. This information appears in the radar return signal as a delay-Doppler shift for the range and range rate respectively. Delay-Doppler shift estimation accuracy is, however, reduced with an increase in the spread of the ambiguity function surface and in the return signal additive noise variance. In non-compressive cognitive radar applications [23–26] the parameters of the transmitted radar waveform are configured to provide a variety of ambiguity function surface landscapes to improve tracking performance based on a prediction on the target state. An example of compressive cognitive radar tracking appears in the application of adaptive compressive sensing and processing (ACSP) [2]. ACSP combines the benefits of compressive sensing, direct processing of the data with no reconstruction, and mitigates the reduction in SNR that results from the use of compressive sensing and processing. Some background on compressive sensing, on direct processing of compressed measurements with no reconstruction, and the use of adaptivity is provided next.

Compressive sensing [71–76] allows signal acquisition at a sub-Nyquist rate while preserving the information contained in the signal. The lower-dimensional signal acquired via compressed sensing also reduces the size of data acquired by the tracking system. Compressive sensing preserves information due to the sparsity of received signals. In the case of target tracking the sparsity in radar signals is attributed to the sparse presence of targets in the field of view of the sensor. Sparsity in the target state space is mapped to sparsity in the delay-Doppler plane that forms the measurement space of the radar sensor [2]. Specifically, the target states are mapped to a set of sparse signal elements, i.e., a small number of elements compared to the dimensionality of the signal that are associated with non-negligible magnitudes. Non-adaptive compressive sensing uses a sensing mechanism that has the benefit of universally preserving information in any sparse signal appearing in a wide range of applications [74]. In order to recover the information preserved in a compressively received signal, reconstruction using l_1-norm optimization [76–81] can be used to extract the sparse elements from the signal.

As an alternative to reconstruction, low-dimensional compressed measurements can be directly processed. Processing compressed measurements is computationally inexpensive compared to processing reconstructed or Nyquist rate measurements. Non-adaptive compressive sensing and processing (NACSP) [82] can then simplify the design of acquisition and processing hardware. The drawback of using NACSP is, however, the reduction of SNR [82] and the increase of delay-Doppler ambiguity function surface sidelobes compared to the use of Nyquist sensing and processing (NSP) [60] which deteriorates tracking performance [54].

Compressive sensing and processing can be improved by using a prediction on the future state of the target at one time step ahead that is mapped to a prediction of the structure of the received signal used to adaptively configure the compressive sensing and processing mechanism. The result of configuration is the improvement of the expected SNR and reduction of the ambiguity function surface sidelobes which improves tracking performance [54, 60, 83]. The adaptive method described in [54, 60, 83] configures the compressive sensing and processing mechanism prior the arrival of new measurements. Another type of adaptive compressive sensing process is described in [84, 85] where signal recovery is improved by configuring the compressive sampling process during signal acquisition.

To summarize the methods described above, compressive sensing achieves signal acquisition at sub-Nyquist rates with the potential to simplify receiver hardware. Moreover, compressively sensed signals can be processed directly to reduce computational load in sequential signal acquisition and processing. NACSP, however, results to a reduction in SNR and deteriorates tracking performance. As a remedy, ACSP utilizes information on the predicted target state to configure the acquisition and processing mechanism in order to achieve an increased predicted

SNR and a more highly peaked ambiguity function surface. The measurement acquisition and processing is next described both at a Nyquist and a compressive rate.

The vector describing the state of the target denoted as $\mathbf{x}_k$ in (3.1) is related to the deterministic version of a Nyquist rate waveform expressed as an M-dimensional vector

$$\mathbf{s}_{l,u} = f'_{u,k}(\mathbf{x}_k). \tag{3.5}$$

Function $f'_{u,k}(\cdot)$ transforms the position and velocity of the target to a delay-Doppler shifted radar signal of length M equal to the length of the radar transmitted sequence with an additional listening time to accommodate for the maximum radar delay with respect to sensor u. Specifically, the state of the target $\mathbf{x}_k$ in (3.1) is converted to range $r_{u,k} = \sqrt{(\chi_u - x_k)^2 + (\psi_u - y_k)^2}$ and range rate $\dot{r}_{u,k} = (\dot{x}_k(x_k - \chi_u) + \dot{y}_k(y_k - \psi_u))/r_{u,k}$, respectively, where χ_u and ψ_u is the location of sensor u. The range and range rate then correspond to delay $\tau_{u,k}$ and Doppler shift $\nu_{u,k}$ where the delay is given as $\tau_{u,k} = \text{round}(\frac{2r_{u,k}/c}{\Delta\tau})$ and the Doppler shift as $\nu_{u,k} = \text{round}(\frac{-2f_c\dot{r}_{u,k}/c}{\Delta\nu})$ where c is the velocity of propagation and f_c is the carrier frequency of the radar signal. Moreover, $\Delta\tau$ and $\Delta\nu$ are the size of the discrete delay and Doppler resolution cells, respectively. The resulting vector $\mathbf{s}_{l,u}$ is indexed by $l_u \in L_{u,k}$ where l_u represents a linear index corresponding to a pair of delay and Doppler shifts of the signal. Function $f'_{\mathrm{s},u,k}(\cdot)$ in radar case does not depend on the set of configuration parameters $\mathbf{b}_{u,k} = \{\mathcal{L}_{\mathrm{b},u,k}\}$ which, in the radar case includes the set of resolution set indices that may contain target returns at the next time step.

Moreover, normalized energy signals $\mathbf{s}_l, l = 1, \ldots, L_u$ form columns of a size $M \times L_u$ matrix $\boldsymbol{\Psi}$. Then, each element is given in terms of $\boldsymbol{\Psi}$ by

$$\mathbf{s}_l = \boldsymbol{\Psi}\boldsymbol{\delta}_l, \tag{3.6}$$

for $l \in \mathcal{L}_u$ where $\boldsymbol{\delta}_l$ is a unit impulse vector with a single non-zero element at index $l \in \mathcal{L}_u$. When Björck Constant Amplitude Zero Autocorrelation Waveforms (CAZACs) are used as radar waveforms the radar waveform satisfies the sparsity requirement [2] which makes $\boldsymbol{\Psi}$ a sparsity dictionary [72]. Björck CAZACs also have a highly peaked ambiguity function surface and low sidelobes which makes CAZACs an attractive choice as radar signals [86, 87]. The compressed C-dimensional counterparts of template vectors given above are given by using (3.6) and a $C \times M$ sensing matrix $\boldsymbol{\Phi}_{\mathrm{b},u,k}$ where $C < M$ as

$$\mathbf{g}_l = \boldsymbol{\Phi}_{\mathrm{b},u,k}\mathbf{s}_l, l \in \mathcal{L}_{\mathrm{b},u,k}. \tag{3.7}$$

In a cognitive setting, $\boldsymbol{\Phi}_{\mathrm{b},u,k}$, has a special structure that will allow its configuration at each time step k of the tracking scenario in order to improve estimation performance. This structure along with the NACSP and NSP sensing matrix structures are provided in Appendix B. The Nyquist rate received signal at radar sensor u is given by

$$\mathbf{r}_{l,u,k} = \gamma_{u,k}\mathbf{s}_{l,u} + \boldsymbol{v}_{u,k}, \tag{3.8}$$

where $\mathbf{r}_{l,u,k}$ is a length M vector that results from sampling the radar return at the Nyquist rate. Moreover, $\gamma_{u,k}$ is a zero-mean complex Gaussian amplitude with variance $2\sigma_\gamma^2$ according to the Swerling I model [70]. $\boldsymbol{v}_{u,k}$ represents length M additive noise vector of zero-mean complex Gaussian i.i.d. elements $v_{u,k}(m), m = 1, \ldots, M$ with variance $2\sigma_v^2$. When using a sub-Nyquist acquisition rate for sampling the radar return the compressed measurement vector is given by

$$\mathbf{h}_{l,u,k} = \mathbf{\Phi}_{\mathrm{b},u,k}\mathbf{r}_{l,u,k}, \tag{3.9}$$

where the acquisition process is represented as a projection of the Nyquist rate signal onto sensing matrix $\mathbf{\Phi}_{\mathrm{b},u,k}$. Compressed vectors $\mathbf{g}_l, l \in \mathcal{L}_{\mathrm{b},u,k}$ given in (3.7) are used for processing compressed measurements in the measurement processing stage described next. The radar waveform acquired compressively in (3.9) is processed at every time step k and for each sensor u by matched filtering with template waveforms given by (3.7) as

$$y_{\mathrm{b},u,k}\left(\bar{l}, l\right) = \mathbf{g}_l^* \mathbf{h}_{\bar{l},u,k} \tag{3.10}$$

for $l \in \mathcal{L}_{\mathrm{b},u,k}$ and the normalized likelihood ratio is given by $\Lambda_{\mathrm{b},\bar{l},l,u,k}$ provided in (2.15) using (2.12), (2.13), and (2.14).

The goal of the cognitive radar process is to increase the expected SNR. The configurable compressive sensing and processing matrix is constructed and configured. The signal to noise ratio at the output of the measurement processing stage is calculated as the ratio of the variance of the signal term to the noise term contained in the matched filter statistic $y_{\mathrm{b},u,k}(l, l)$ in (3.8) with $\bar{l} = l$ and using (3.6), (3.7), and (3.9) the SNR is given by

$$\eta_{l,l,u,k} = \frac{\sigma_\gamma^2}{\sigma_v^2} \frac{|\mathbf{g}_l|^4}{\mathbf{g}_l^H \mathbf{\Phi}_{\mathrm{b},u,k} \mathbf{\Phi}_{\mathrm{b},u,k}^H \mathbf{g}_l}. \tag{3.11}$$

The SNR at the input of the processing stage, on the other hand, is given by

$$R_{IP} = \frac{\sigma_\gamma^2 \xi_s}{\sigma_v^2}. \tag{3.12}$$

The expected SNR is estimated next in order to enable sensor configuration as described in Chapter 2. The expected SNR is estimated based on the estimate of the probability distribution of resolution cells that the true state is likely to being mapped to for sensor u at one time step ahead given by (2.27). From (3.11) the expected SNR for each node u and time step k is given by

$$\bar{\eta}_{l,l,u,k} = \frac{\sigma_\gamma^2}{\sigma_v^2} \sum_{l \in \mathcal{L}_{\mathrm{b},u,k}} \frac{|\mathbf{g}_l|^4}{\mathbf{g}_l^H \mathbf{\Phi}_{\mathrm{b},u,k} \mathbf{\Phi}_{\mathrm{b},u,k}^H \mathbf{g}_l} p_{\mathrm{b},u,k}(l) \tag{3.13}$$

which is taken as the expected value of the SNR expression in (3.11) over indices $l \in \mathcal{L}_{\mathrm{b},u,k}$. The parameters that need to be configured so that the expected SNR for index l can change value for different probabilities $p_{\mathrm{b},u,k}(l)$ are related to matrix $\mathbf{\Phi}_{\mathrm{b},u,k}$ used in sensing and processing provided in Appendix B where different configurations of the matrix are outlined that consider available information, partial available information, or no information on the sequentially updated target state.

3.2 TARGET TRACKING METHOD

The sequential estimation method for tracking a single target is described next. The method samples hypotheses on the target state from the target state evolution distribution. The hypotheses are mapped from the target state space to the measurement space of each node to deliver a probability distribution of a set of the resolution cells onto which the target state will likely map to at the next time step. The probability distribution is used at each time step to (a) select the angle of gaze and the subset of angular resolution cells to activate for sensing and processing at the foveal node and (b) select the composition of the compressive radar sensing matrix to improve the expected SNR for compressive cognitive radar nodes. Node configuration is followed by measurement acquisition, and an auxiliary particle sampling process that generates particles that represent both prior information and information from measurements to obtain a discrete representation of the posterior distribution estimate.

3.2.1 PREDICTION

The posterior of the state of the target is evaluated at discrete locations $\mathbf{x}_{n,k},\ n = 1,\ldots,N$ which are associated with weights representing probability values. Initially, predicted states are generated using the state evolution model in (3.1) as

$$\mathbf{x}_{n,k} = \mathbf{F}\mathbf{x}_{n,k-1} + \mathbf{V}\boldsymbol{\eta}_{n,k}, n = 1,\ldots,N. \tag{3.14}$$

Predicted states are then mapped to the measurement space of each sensor $u = 1,\ldots,U$ resulting to template vectors

$$\mathbf{s}_{l,u,k} = f'_{\mathrm{b},u,k}\left(\mathbf{x}_{n,k}\right) \tag{3.15}$$

for $n = 1,\ldots,N, u = 1,\ldots,U$ as in (3.3) and (3.5) for the foveal and radar node, respectively, that can be used for measurement processing and that are indexed by

$$l \in \mathcal{L}'_{\mathrm{b},u,k} = \bigcup \{l_{n,u}\}_{n=1}^{N} \tag{3.16}$$

as in (2.26) from which a connected set $\mathcal{L}_{\mathrm{b},u,k}$ is generated via interpolation from the possibly unconnected set $\mathcal{L}'_{\mathrm{b},u,k}$.

Next, the probability associated with the appearance of signal elements indexed by $l \in \mathcal{L}_{\mathrm{b},u,k}$ in future measurements is calculated based on the prior information available from the tracking process as

$$p_{\mathrm{b},u,k}(l) = \sum_{n:\mathbf{s}_{l,u,k}=f'_{\mathrm{b},u,k}(\mathbf{x}_{n,k})} w_{n,k-1} \tag{3.17}$$

$u = 1, \dots, U$, $l \in \mathcal{L}_{b,u,k}$ as in (2.27). The set $w_{n,k-1}$ represents particle weights at the previous time step $k-1$. Particle weights were defined in (2.31). The set $\mathcal{L}'_{b,u,k}$ then includes signal elements in the respective measurement spaces of individual sensors that may appear in the received signal at the next time step with non-negligible probability $p_{b,u,k}(l)$ associated with generated hypotheses on the target state $\mathbf{x}_{n,k}, n = 1, \dots, N$. For the foveal nodes the configuration parameter that affects $p_{b,u,k}(l)$ is the angle of gaze $\theta_{o,u,k}$ and for the radar nodes $p_{b,u,k}(l)$ is not affected by configuration parameters.

3.2.2 FOVEAL NODE CONFIGURATION

Using (3.2) with $\theta_{o,u,k} = 0$, predicted states map onto predicted angles

$$\vartheta'_{n,u,k} = \arctan \frac{\psi_{n,k} - y_u}{\chi_{n,k} - x_u}, n = 1, \dots, N \tag{3.18}$$

where $\vartheta'_{n,u,k}$ is the angle where the target will appear with respect to foveal sensor u according to the hypothesis represented by particle n. A prediction of the estimated angle formed by the target and foveal node u is next obtained as

$$\theta_{o,u,k} = \sum_{n=1}^{N} w_{n,k-1} \vartheta'_{n,u,k}, \tag{3.19}$$

where $w_{n,k-1}$ represents particle weights estimated at $k-1$ defined in the next subsection. Angle $\theta_{o,u,k}$ is then set to be the configured angle of gaze of the node.
where $w_{n,k-1}$ represents particle weights estimated at time step $k-1$ and defined in the next subsection. $\theta_{o,u,k}$ is then the configured angle of gaze. The set of angles associated with each particle after node configuration are

$$\vartheta_{n,u,k} = \arctan \frac{\psi_{n,k} - y_u}{\chi_{n,k} - x_u} - \theta_{o,u,k}, n = 1, \dots, N \tag{3.20}$$

with resolution cell indices

$$l_{n,u} = \underset{l}{\operatorname{argmin}} \, |\vartheta_{n,u,k} - \theta_{l,u}|, \; n = 1, \dots, N. \tag{3.21}$$

Moreover, for both the foveal and the radar node

$$l \in \mathcal{L}_{b,u,k} = \bigcup \{l_{n,u}\}_{n=1}^{N} \tag{3.22}$$

represents indices of resolution cells that are associated with the true target state at one step ahead with probability

$$p_{b,u,k}(l) = \sum_{n: s_{l,u,k} = f_{b,u,k}(\mathbf{x}_{n,k})} w_{n,k-1}, \; l \in \mathcal{L}_{b,u,k}, \tag{3.23}$$

where $f_{\mathrm{b},u,k}(\mathbf{x}_{n,k})$ is defined in (3.3) and (3.5) for foveal and radar nodes, respectively. Indices with zero or negligible probability of representing the target state will not be sampled and $\mathcal{L}_{\mathrm{b},u,k} \subset \mathcal{L}_u$, thus preserving resources in all nodes.

3.2.3 RADAR NODE CONFIGURATION

The cognitive radar nodes are configured by adapting the sensing matrix in order to increase the expected SNR based on the probability of signal elements to appear in the measurements given by $p_{\mathrm{b},u,,k}(l)$. Different structures for the cognitive sensing matrix $\mathbf{\Phi}_{\mathrm{b},u,k}$ can be used as explained in [2] where all structures utilize the probability of signal elements indexed by l given by $p_{\mathrm{b},u,k}(l)$ to appear in the measurements. Moreover, the number of measurements C to be collected can be varied as a configuration parameter. An increase in C results to an increase in sensing and processing resources providing the benefit of an improved SNR. Appendix B explains the configuration of the sensing matrix for different ACSP methods.

3.2.4 MEASUREMENT UPDATE

Sensing and processing by the angular and radar nodes results to measurement statistics that are communicated to the fusion center. After the calculation of the likelihood ratio corresponding to each node given in (2.15) using (2.12), (2.13), and (2.14) an index is sampled from the likelihood ratio for each particle n and for each sensor u as

$$\tilde{l}_{n,u} \sim \left\{\Lambda_{\mathrm{b},\bar{l},l,u,k}\right\}_{l\in\mathcal{L}_{\mathrm{b},u,k}} \tag{3.24}$$

as in (2.28) with sampling bias $\tilde{\Lambda}_{\mathrm{b},n,u,k}$ where $l = \tilde{l}_{n,u}$. Next, a process of auxiliary sampling using prior information takes place for each particle $\mathbf{x}_{n,k-1}, n = 1, \ldots, N$

$$\acute{\mathbf{x}}_{n',k} = \mathbf{F}\mathbf{x}_{n,k-1} + \mathbf{V}\boldsymbol{\eta}_{n',k} \tag{3.25}$$

for $n' = 1, \ldots, N'$ similarly to (3.14). Using (3.25) with (3.15) as $\mathbf{s}_{l',u,k} = f'_{\mathrm{b,s},u}(\acute{\mathbf{x}}_{n',k})$ then $\acute{\mathbf{x}}_{n',k}$ is mapped to index $l = l_{n',u}$. Then a sample n'' out of $n' = 1, \ldots, N'$ sampled particles are identified such that

$$n'' = \underset{n'}{\operatorname{argmin}} \left\| l_{n',u} - \tilde{l}_{n,u} \right\| \tag{3.26}$$

for sensors $u = 1, \ldots, U$ to select states of targets $\mathbf{x}_{n,k} = \acute{\mathbf{x}}_{n'',k}$ for $n = 1, \ldots, N$ that correspond to measurement sampled indices.

Particle Weights and Estimation

Particles $\acute{\mathbf{x}}_{n',k}, n' = 1, \ldots, N'$, sampled in (3.25) receive weights [48] based on how well they represent prior information and measurement information, also taking into account the bias with which they have been sampled. The weights expression that includes the sampling bias is

$$w_{n,k} = \frac{p\left(\acute{\mathbf{x}}_{n',k}|\mathbf{x}_{n,k-1}\right)\prod_{u=1}^{U}\Lambda_{\bar{l},l,u,k}}{p\left(\acute{\mathbf{x}}_{n',k}|\mathbf{x}_{n,k-1}\right)\prod_{u=1}^{U}\tilde{\Lambda}_{\bar{l},l,u,k}}, \quad n' = 1,\ldots,N', \tag{3.27}$$

where the numerator represents the prior distribution evaluated at $\mathbf{x}_{n,k}$ and the likelihood distribution in (2.15) $l = l_{n,u}$ where $l_{n,u}$ is the index on which $\mathbf{x}_{n,k}$ is mapped onto in the measurement space of each node u. Moreover, biases in the denominator are due to sampling particle $\mathbf{x}_{n,k}$ from the prior at (3.25) and sampling an index $l = \tilde{l}_{n,u}$ from the likelihood in (3.24). The prior probabilities in the numerator and denominator cancel out. Moreover, if $l_{n,u} = \tilde{l}_{n,u}$then the likelihood terms cancel out with $w_{n,k} = 1$. Otherwise, the weights become

$$w_{n,k} = \prod_{u=1}^{U}\frac{\Lambda_{\bar{l},l,u,k}}{\tilde{\Lambda}_{\bar{l},l,u,k}} \tag{3.28}$$

for each particle $n = 1,\ldots,N$.

Since all particles are sampled using both the prior and likelihood then particle weights [48] are assigned as $w_{n,k} = w_{n,k-1}$ for $n = 1,\ldots,N$ if $l_{n,u} = \tilde{l}_{n,u}$ and zero otherwise for all sensors u. The estimated posterior is

$$p\left(\hat{\mathbf{x}}_k|\mathbf{y}_k\right) = \sum_{n=1}^{N} w_{n,k}\delta\left(\mathbf{x}_k - \mathbf{x}_{n,k}\right) \tag{3.29}$$

and the MMSE estimate is given by

$$\hat{\mathbf{x}}_k = \sum_{n=1}^{N} w_{n,k}\mathbf{x}_{n,k}. \tag{3.30}$$

A resampling process follows to avoid degeneracy [48].

3.3 SIMULATION SCENARIO

A simulation-based study [64] is provided to demonstrate tracking using heterogeneous nodes. The method is shown to have the ability to improve tracking performance via the fusion of heterogeneous data and preserve node resources via adaptively allocating resources based on sequentially updated information on the target state. To demonstrate the benefits of cognitive heterogeneous data fusion simulations were repeated using different configurations of the sequential Monte Carlo method described in this chapter using one vs. two types of nodes and fixed vs. adaptive configurations. Specifically, cognitive foveal nodes and fixed configuration angle nodes were used having fixed angular width resolution cells. Moreover, cognitive radar nodes and fixed sampling rate radar nodes were used. The simulation results evaluate the performance of tracking a single target where the tracking performance is quantified by the root mean squared error (RMSE) and percentage of lost tracks.

In the simulation-based study, a single target moves on a two-dimensional Cartesian plane, where a typical trajectory is shown in Figure 3.2. The experiments described were repeated over 100 Monte Carlo runs, where a new random trajectory was generated for each Monte Carlo run. The target motion is completed in 98 time steps. The target motion is observed by a network of radar and angular sensing nodes which collect one measurement vector at each time step. Specifically, 3 foveal nodes are located at $x = 1000, y = 500, x = 2500, y = 0$, and $x = 4000, y = 500$ and 3 radar nodes are located at $x = 2000, y = 0, x = 3500, y = 0$, and $x = 500, y = 0$. The SNR for both sensors was varied to take values of $14, 16, 18, 20$ dB, respectively.

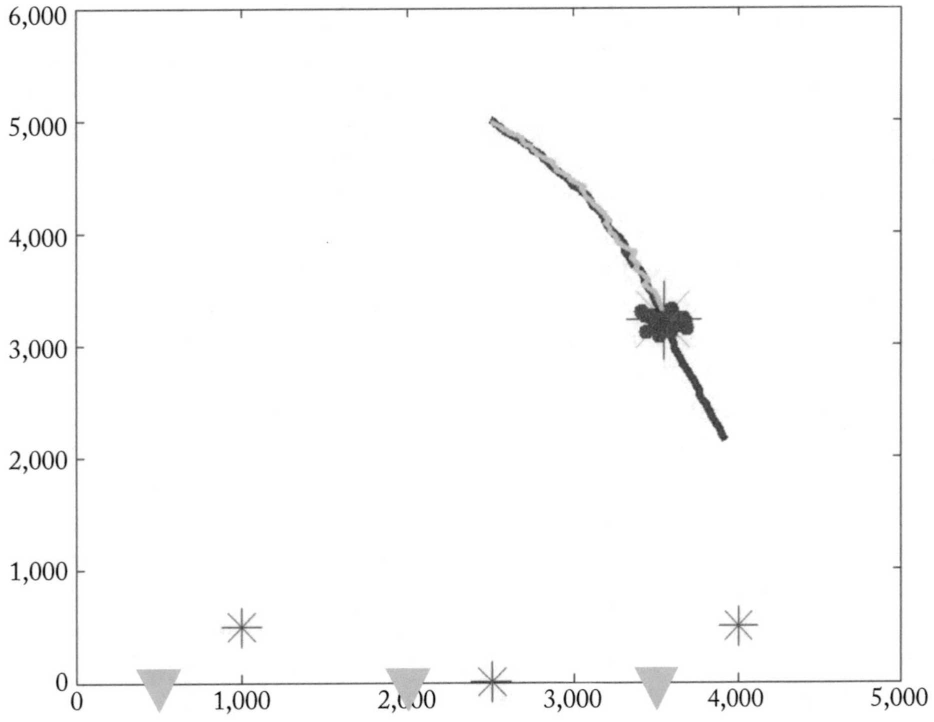

Figure 3.2: Simulation setup including a typical target trajectory (blue) and estimated trajectory (green). The position of three radar sensors and three foveal/angle sensors is indicated with a '▼' and '*', respectively.

Four different setups are considered for the angular sensor which covered a range of π rad. The first two were fixed, having high and low resolution, respectively. In the high-resolution setup the angular sensor had $L_{b,u,k} = 315$ cells of equal angle 0.01 rad and in the low-resolution

$L_{b,u,k} = 126$ of equal angle 0.025 rad. The third setup was a foveal sensing case where the central foveal region is composed of in total 187 cells having 126 cells of equal angle 0.001 rad in the central high-resolution foveal region, 52 cells in a lower resolution peripheral region of equal angle 0.01 rad, and 9 cells of lowest angle 0.1. Therefore, in the foveal sensing case the savings result from sensing and processing less resolution cells than in the high resolution case. Moreover, the foveal sensing case affords higher resolution in the foveal region compared to the high resolution case with less number of cells overall.

Considering to the radar sensor, the radar transmitted waveform was a Björck CAZAC [87], [86] sequence of prime number length 9973. The dimensionality of the return signal received at the radar sensor using a Nyquist rate sampling was set to $M = 11{,}000$ to include a sufficiently high time delay in the return signal. Radar signals were transmitted with center frequency of 40 GHz and the Nyquist sampling rate was set to 10 MHz. For the adaptive sampling CSP (ASCSP) and NACSP the number of compressive measurements was set to $C = 3000$ and the dictionary size was fixed to $L_{b,u,k} = 4000$ which was found enough to describe all the fluctuating uncertainty of the target so that different options of the matrices composing the sensing matrix are pre-calculated. The resource use of the radar node is measured in terms of the length of the acquired signal. In the non-cognitive NSP this number is the dimensionality of the Nyquist rate sampled signal $M = 11{,}000$, while in the non-cognitive and cognitive compressive sensing methods this number is the number of compressive measurements $C = 3000$ which is much smaller than the dimensionality of the Nyquist rate signal.

In Figures 3.3 and 3.4, the RMSE provided with 95 percent confidence intervals and the percentage of lost tracks are plotted, respectively, vs. the SNR where a lost track is counted if the estimated state deviates for more than 500 m from the true state for more than 5 consecutive time steps. The following configurations were used for the simulation scenarios: (i) High-NSP Fusion: using both high-resolution angular and NSP radar nodes; (ii) NSP Radar: using only NSP radar nodes; (iii) High Angular: using only angular nodes with high resolution; (iv) Low-NACSP Fusion: using both low resolution angular and radar NACSP; and (v) Cognitive Fusion: using both foveal and radar with ASCSP. The results show that when fusing data from two types of nodes the performance is consistently better than when using one type of node. In addition, both the adaptive configuration and high-resolution angular and NSP radar provide an improved performance vs. a fusion of lower resolution angular and NACSP radar. In addition, when using an adaptive configuration vs. a fixed configuration with high-resolution settings the performance is similar. However, the adaptive configuration by the cognitive system configuration delivers high performance with lower resource use compared to the fixed configuration case using high-resolution angular and NSP radar.

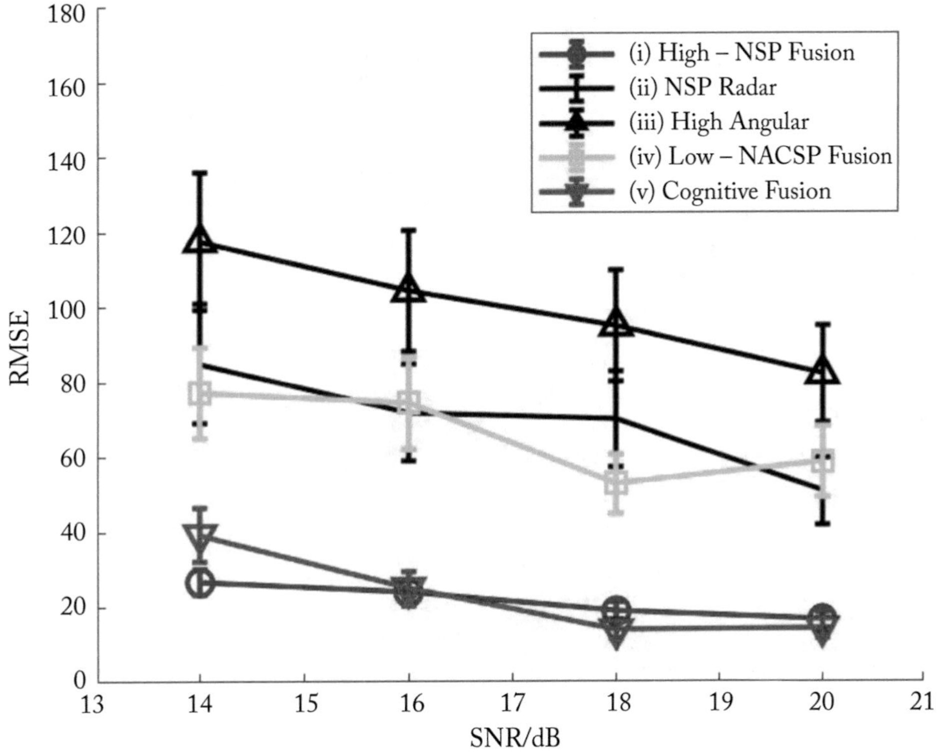

Figure 3.3: RMSE tracking error performance vs. the SNR.

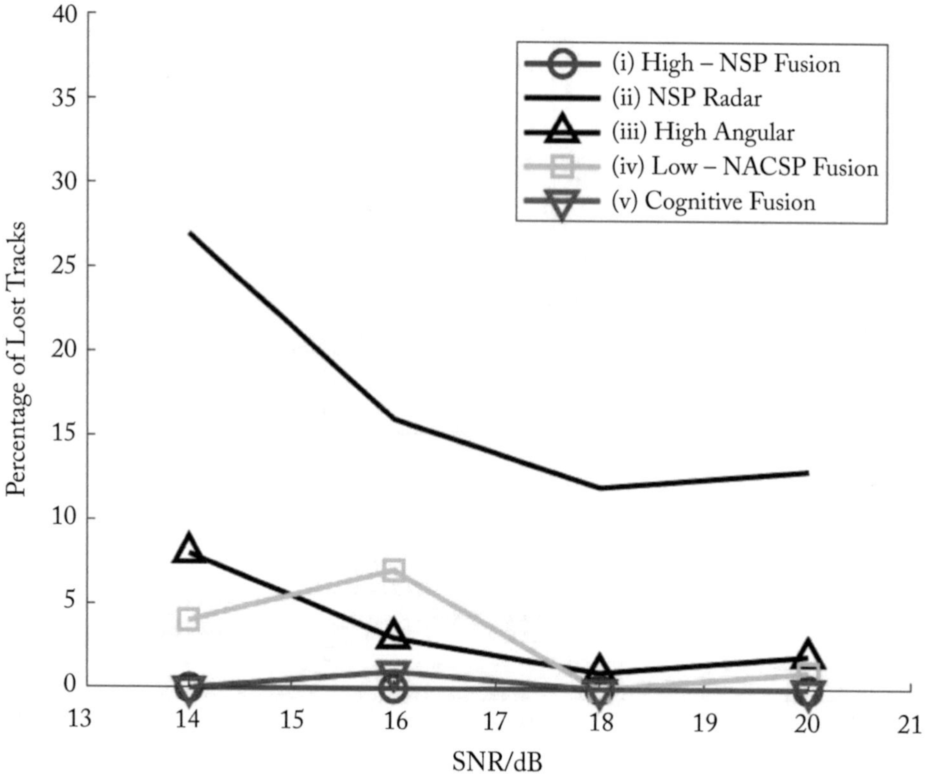

Figure 3.4: Number of lost tracks vs. the SNR.

CHAPTER 4

Conclusions

Configurable heterogeneous nodes provide a diversity of information on the estimated state and adaptively allocate their resources to improve information acquisition. Therefore, the availability of multiple heterogeneous sensing nodes that have the ability to be reconfigured in terms of sensing, processing, and communications presents the opportunity to improve estimation performance.

However, the processes of fusing heterogeneous data and adapting configuration parameters of heterogeneous sensing nodes is associated with a number of challenges. These challenges are due to the often nonlinear relationship between the state and measurements and the diversity of the relationships between state and measurements from nodes of different types and resolution capabilities. Moreover, sequential estimation of the possibly highly non-Gaussian posterior distribution of the state cannot be efficiently handled with simplifying assumptions of linearity and Gausianity. The problem of configuring each node one time step ahead is also challenging due to the need to utilize information from a fusion of data from all nodes to configure the particular settings of each individual node.

This work has described cognitive fusion, a general framework for heterogeneous data fusion and node configuration that is based on the Bayesian inference that addresses the above challenges and benefits from the use of configurable heterogeneous nodes. The method solves the dual problem of heterogeneous sensing node data fusion and adaptive configuration of heterogeneous nodes. Moreover, it handles realistic measurement models with no simplifying assumptions. Specifically, the method sequentially updates information on the state that is captured in the estimated posterior distribution of the state and then utilizes the information to configure sensing nodes at the next time step. A simulation-based study was provided to demonstrate a practical application of cognitive fusion using configurable angular and radar sensing nodes. In the simulation study a single target is tracked using radar sensing nodes delivering either compressive or Nyquist rate measurements and angular sensors of varying resolution capabilities. The radar sensing node is equipped with adaptive compressive sensing capabilities, while the angular sensor is capable of adaptive foveal sensing. The results of the simulation-based study illustrate the ability of the method to improve tracking by using configurable heterogeneous nodes, therefore, providing a practical example of cognitive fusion for target tracking. To provide descriptions of various methods for cognitive fusion, this book also cites several references. Suggested further reading in compressive sensing includes [71–74]. Applications are covered in [46, 54, 60, 75–78], [84, 85], and [88, 89] and a software tool is provided in [79]. Moreover, futher reading on sensor fusion and localization is found in [90], [91].

APPENDIX A

Sensing Node Configuration

The MSE is defined in Chapter 2 in (2.18) as

$$\mathcal{E}_{\mathbf{B},k} = E_{\mathbf{B},k}\left[(\bar{\mathbf{x}}_k - \mathbf{x}_k)^T \mathbf{C}(\bar{\mathbf{x}}_k - \mathbf{x}_k)\right]. \quad \text{(A.1)}$$

The MSE is, furthermore, explicitly written as the expectation over the distribution of the true and estimated states as

$$\mathcal{E}_{\mathbf{B},k} = \int_{\bar{\mathbf{x}}_k}\int_{\mathbf{x}_k} (\bar{\mathbf{x}}_k - \mathbf{x}_k)^T \mathbf{C}(\bar{\mathbf{x}}_k - \mathbf{x}_k)\, p\left(\bar{\mathbf{x}}_k, \mathbf{x}_k | \mathbf{Y}_{\mathbf{B},k}, \mathbf{Y}_{k-1}\right) d\mathbf{x}_k d\bar{\mathbf{x}}_k. \quad \text{(A.2)}$$

The joint probability $p(\bar{\mathbf{x}}_k, \mathbf{x}_k | \mathbf{Y}_{\mathbf{B},k}, \mathbf{Y}_{k-1})$ is expressed as

$$p\left(\bar{\mathbf{x}}_k, \mathbf{x}_k | \mathbf{Y}_{\mathbf{B},k}, \mathbf{Y}_{k-1}\right) = p\left(\mathbf{x}_k | \bar{\mathbf{x}}_k, \mathbf{Y}_{\mathbf{B},k}, \mathbf{Y}_{k-1}\right) p\left(\bar{\mathbf{x}}_k\right). \quad \text{(A.3)}$$

Moreover, [48]

$$p\left(\mathbf{x}_k | \bar{\mathbf{x}}_k, \mathbf{Y}_{\mathbf{B},k}, \mathbf{Y}_{k-1}\right) = \frac{p\left(\mathbf{Y}_{\mathbf{B},k} | \mathbf{x}_k, \bar{\mathbf{x}}_k\right) p\left(\mathbf{x}_k | \bar{\mathbf{x}}_k, \mathbf{Y}_{k-1}\right)}{\int_{\mathbf{x}_k} p\left(\mathbf{Y}_{\mathbf{B},k} | \mathbf{x}_k, \bar{\mathbf{x}}_k\right) p\left(\mathbf{x}_k | \bar{\mathbf{x}}_k, \mathbf{Y}_{k-1}\right) d\mathbf{x}_k} \quad \text{(A.4)}$$

and using (2.11), (2.14), and (2.15) together with the fact that likelihoods related to nodes other than u are not affected when changing configuration settings of node u then the following proportionality results from (A.4):

$$p\left(\mathbf{x}_k | \bar{\mathbf{x}}_k, \mathbf{Y}_{\mathbf{B},k}, \mathbf{Y}_{k-1}\right) \propto \Lambda_{\mathrm{b},\bar{l},l,u,k}\, p\left(\mathbf{x}_k | \bar{\mathbf{x}}_k, \mathbf{Y}_{k-1}\right), u = 1, \ldots, U. \quad \text{(A.5)}$$

Moreover, taking into account that states $\bar{\mathbf{x}}_k : \mathbf{s}_{\bar{l},u,k} = f'_{\mathrm{b},u,k}(\bar{\mathbf{x}}_k)$, and $\mathbf{x}_k : \mathbf{s}_{l,u,k} = f'_{\mathrm{b},u,k}(\mathbf{x}_k)$ are mapped, using (2.3), to resolution cells of node u indexed by $\bar{l}$ and l and by replacing (A.5) and (A.3) into (A.2) yields

$$\mathcal{E}_{\mathbf{B},k} \propto \sum_{\bar{l}\in\mathcal{L}_{\mathrm{b},u,k}} \sum_{l\in\mathcal{L}_{\mathrm{b},u,k}} \Lambda_{\mathrm{b},\bar{l},l,u,k} \int_{\bar{\mathbf{x}}_k : \mathbf{s}_{\bar{l},u,k} = f'_{\mathrm{b},u,k}(\bar{\mathbf{x}}_k)} \int_{\mathbf{x}_k : \mathbf{s}_{l,u,k} = f'_{\mathrm{b},u,k}(\mathbf{x}_k)} (\bar{\mathbf{x}}_k - \mathbf{x}_k)^T \mathbf{C}(\bar{\mathbf{x}}_k - \mathbf{x}_k)\, p\left(\mathbf{x}_k | \bar{\mathbf{x}}_k, \mathbf{Y}_{k-1}\right) p\left(\bar{\mathbf{x}}_k\right) d\mathbf{x}_k d\bar{\mathbf{x}}_k \quad \text{(A.6)}$$

for each $u = 1, \ldots, U$ where the term in the error expression when states are grouped within sets of states that fall inside resolution cells indexed by l and $\bar{l}$ is given by

$$\varepsilon_{\mathrm{b},\bar{l},l,u,k} = \int_{\bar{\mathbf{x}}_k : \mathbf{s}_{\bar{l},u,k} = f'_{\mathrm{b},u,k}(\bar{\mathbf{x}}_k)} \int_{\mathbf{x}_k : \mathbf{s}_{l,u,k} = f'_{\mathrm{b},u,k}(\mathbf{x}_k)} (\bar{\mathbf{x}}_k - \mathbf{x}_k)^T \mathbf{C} (\bar{\mathbf{x}}_k - \mathbf{x}_k)$$
$$p(\mathbf{x}_k | \bar{\mathbf{x}}_k, \mathbf{Y}_{k-1}) \, p(\bar{\mathbf{x}}_k) \, d\mathbf{x}_k d\bar{\mathbf{x}}_k. \quad \text{(A.7)}$$

The probability that a target state will map onto a specific resolution cell is then calculated using (2.4) as $p_{\mathrm{b},u,k}(\bar{l}) = \int_{\bar{\mathbf{x}}_k : \mathbf{s}_{\bar{l},u,k} = f'_{\mathrm{b},u,k}(\bar{\mathbf{x}}_k)} p(\bar{\mathbf{x}}_k) d\bar{\mathbf{x}}_k$ and $p_{\mathrm{b},u,k}(l) = \int_{\mathbf{x}_k : \mathbf{s}_{l,u,k} = f'_{\mathrm{b},u,k}(\mathbf{x}_k)} p(\mathbf{x}_k | \bar{\mathbf{x}}_k, \mathbf{Y}_{k-1}) d\mathbf{x}_k$. Moreover, it is assumed that the distributions $p(\bar{\mathbf{x}}_k)$ and $p(\mathbf{x}_k | \bar{\mathbf{x}}_k, \mathbf{Y}_{k-1})$ are uniform over the relatively small sized resolution cells and their values are equal to $p_{\mathrm{b},u,k}(\bar{l})$ and $p_{\mathrm{b},u,k}(l)$, respectively. Then from (A.7) $\epsilon_{\mathrm{b},\bar{l},l,u,k}$ is defined as the resolution cell error

$$\epsilon_{\mathrm{b},\bar{l},l,u,k} = \int_{\bar{\mathbf{x}}_k : \mathbf{s}_{\bar{l},u,k} = f'_{\mathrm{b},u,k}(\bar{\mathbf{x}}_k)} \int_{\mathbf{x}_k : \mathbf{s}_{l,u,k} = f'_{\mathrm{b},u,k}(\mathbf{x}_k)} (\bar{\mathbf{x}}_k - \mathbf{x}_k)^T \mathbf{C} (\bar{\mathbf{x}}_k - \mathbf{x}_k) \, d\mathbf{x}_k d\bar{\mathbf{x}}_k. \quad \text{(A.8)}$$

Using (A.8) and (2.4) the error in (A.6) then becomes

$$\mathcal{E}_{\mathrm{B},k} \propto \sum_{\bar{l} \in \mathcal{L}_{u,k}} \sum_{l \in \mathcal{L}_{u,k}} \Lambda_{\mathrm{b},\bar{l},l,u,k} \epsilon_{\mathrm{b},\bar{l},l,u,k} \, p_{\mathrm{b},u,k}(\bar{l}) \, p_{\mathrm{b},u,k}(l), u = 1, \ldots, U. \quad \text{(A.9)}$$

The additive terms in the MSE expression are next separated in two sets. The first includes values of state $\mathbf{x}_k$ and true state $\bar{\mathbf{x}}_k$ that map onto the same resolution cell (i.e., $\bar{\mathbf{x}}_k : \mathbf{s}_{\bar{l},u,k} = f'_{\mathrm{b},u,k}(\bar{\mathbf{x}}_k)$, $\mathbf{x}_k : \mathbf{s}_{l,u,k} = f'_{\mathrm{b},u,k}(\mathbf{x}_k), \bar{l} = l$) while the second integral includes all other values that $\mathbf{x}_k$ can take. (A.9) becomes

$$\mathcal{E}_{\mathrm{B},k} \propto \sum_{l \in \mathcal{L}_{u,k}} \Lambda_{\mathrm{b},l,l,u,k} \epsilon_{\mathrm{b},l,l,u,k} \, p^2_{\mathrm{b},u,k}(l)$$
$$+ \sum_{\bar{l} \in \mathcal{L}_{u,k}} \sum_{l \in \mathcal{L}_{u,k}, l \neq \bar{l}} \Lambda_{\mathrm{b},\bar{l},l,u,k} \epsilon_{\mathrm{b},\bar{l},l,u,k} \, p_{\mathrm{b},u,k}(\bar{l}) p_{\mathrm{b},u,k}(l), u = 1, \ldots, U. \quad \text{(A.10)}$$

The error between two states that lie in different resolution cells is generally larger than the error between two states that lie in the same resolution cell ($\epsilon_{\mathrm{b},l,l,u,k} < \epsilon_{\mathrm{b},\bar{l},l,u,k}$). Therefore, to reduce the overall error, resolution cells l associated with a higher probability $p_{\mathrm{b},u,k}(l)$ should be associated with higher likelihood ratio values $\Lambda_{\mathrm{b},l,l,u,k}$ and lower-resolution cell error $\epsilon_{\mathrm{b},l,l,u,k}$. Moreover, the likelihood ratio is proportional to the SNR and the ambiguity function mainlobe. Therefore, from the expression in (A.10) suboptimal computationally inexpensive goals are derived that reduce the MSE as described in Chapter 2 which are based on increasing the expected SNR and reducing the expected ambiguity function sidelobe level and expected resolution cell error.

APPENDIX B

Adaptive Compressive Sensing Matrix

B.1 SENSING MATRIX CONSTRUCTION

An improvement in expected SNR is accomplished by adapting the signal acquisition and processing method at one time step ahead based on information on the state available from the sequential estimation process. The SNR depends on the sensing matrix $\mathbf{\Phi}_{u,k}$ which needs to become configurable and be adapted at every time step. The construction of a configurable sensing matrix for ACSP and the non-adaptive NACSP and NSP sensing matrices are presented in this section [2, 83]. The configurable sensing matrix structure in ACSP is given by

$$\mathbf{\Phi}_{u,k} = \mathbf{QAS}^H \tag{B.1}$$

that also maintains the properties of sparsity and incoherence (see [2, 83] for more detail). In (B.1), matrix $\mathbf{Q}$ is composed by rows that form an orthonormal set and by columns that have approximately constant inner products $\mathbf{q}_l^H \mathbf{q}_l, l = 1, \ldots, L_{\mathrm{b},u,k}$ having the restricted isometry property [2, 82]. $\mathbf{A}$ is an $L_{\mathrm{b},u,k} \times L_{\mathrm{b},u,k}$ diagonal matrix that contains elements $\alpha_l, l \in \mathcal{L}_{\mathrm{b},u,k}$ in its main diagonal that act as weights of individual signal elements and form a configuration parameter of the sensing matrix. Moreover, matrix $\mathbf{S}$ is constructed with template signals $\mathbf{s}_l$ in its columns with $l \in \mathcal{L}_{\mathrm{b},u,k}$ and is an M by $L_{\mathrm{b},u,k}$ matrix. Matrix $\mathbf{S}$ is then configured at every time step with a selected number of elements using the time varying set $\mathcal{L}_{\mathrm{b},u,k}$ containing information on the future measurement structure based on information on the future state of the target. The number of measurements C can also be configured depending on the particular tracking scenario, where increasing C improves SNR and tracking performance [54, 60, 83]. Using the sensing matrix construction in (B.1) the expected SNR is given in (3.13) is now given for sensor u and at time step k as

$$\bar{\eta}_{l,l,u,k} = \frac{\sigma_\gamma^2}{\sigma_\upsilon^2} \sum_{l=1}^{L_{\mathrm{b},u,k}} \frac{|\mathbf{g}_l|^4}{\mathbf{g}_l^H \mathbf{\Phi}_{u,k} \mathbf{\Phi}_{u,k}^H \mathbf{g}_l} p_{\mathrm{b},u,k}(l). \tag{B.2}$$

It can be shown (see [2] for more detail) that when using the adaptive matrix construction in (B.1) the expected SNR expression becomes

$$\bar{\eta}_{l,l,u,k} = \frac{\sigma_\gamma^2}{\sigma_v^2} \frac{C^2}{L_{\mathrm{b},u,k}^2} \sum_{l=1}^{L_{\mathrm{b},u,k}} \frac{\alpha_l^2}{\mathbf{q}_l^H \mathbf{Q}\mathbf{A}\mathbf{A}^H \mathbf{Q}^H \mathbf{q}_l} p_{\mathrm{b},u,k}(l). \tag{B.3}$$

B.2 SENSING MATRIX CONFIGURATIONS

ACSP takes different forms based on the level of reliance on prior information [2]. The ASCSP considers the probability of the true state mapping onto resolution cell l by sampling a subset of the non-negligible probability signal elements. Sampling signal elements reduces dimensionality and simplifies sensing and processing. The NACSP matrix [82] is provided as a special case of the adaptive matrix that ignores all prior information which makes the method universally applicable. However, the use of NACSP reduces SNR and deteriorates tracking performance. The NSP is also presented which, similarly to the NACSP ignores prior tracking information, however, collects a number of measurements equal to the dimensionality of the received signal achieving the best tracking performance among all matrices at the cost of higher sensing and processing load. The expected SNR for each of the sensing and processing methods is provided next.

B.2.1 ADAPTIVE SAMPLING CSP (ASCSP)

ASCSP utilizes prior information from the tracking process by sampling from the probability distribution of the true state mapping onto each resolution cell l, $p_{\mathrm{b},u,k}(l)$, $l = 1, \ldots, L_{\mathrm{b},u,k}$. Signal elements from $\mathcal{L}_{\mathrm{b},u,k}$ are then selected via sampling with no replacement to create a smaller set $\tilde{\mathcal{L}}_{\mathrm{b},u,k}$ with cardinality $\tilde{L}_{\mathrm{b},u,k}$. Since the probability distribution of signal elements in (2.27) has been already utilized in the sampling step, matrix $\mathbf{A}$ in (B.1) becomes an $\tilde{L}_{\mathrm{b},u,k} \times \tilde{L}_{\mathrm{b},u,k}$ identity matrix. In addition, matrix $\mathbf{S}^H$ is constructed based on the sampled indices $l \in \tilde{\mathcal{L}}_{\mathrm{b},u,k}$ and $\mathbf{Q}$ is a random $C \times \tilde{L}_{\mathrm{b},u,k}$ matrix with orthonormal rows and columns $\mathbf{q}_l$ $l = 1, \ldots, \tilde{L}_{\mathrm{b},u,k}$. The sensing matrix then becomes $\mathbf{\Phi}_{\mathrm{b},u,k} = \mathbf{Q}\mathbf{S}^H$. The resulting expected SNR expression in (B.3) using the construction described above becomes

$$\bar{\eta}_{l,l,u,k} = \frac{\sigma_\gamma^2}{\sigma_v^2} \frac{C}{\tilde{L}_{\mathrm{b},u,k}} \sum_{l=1}^{\tilde{L}_{\mathrm{b},u,k}} p_{\mathrm{b},u,k}(l). \tag{B.4}$$

From the expected SNR expression it can be seen that reducing $\tilde{L}_{\mathrm{b},u,k}$ improves the expected SNR. However, reducing the value of $\tilde{L}_{\mathrm{b},u,k}$ increases the probability that a signal element appearing in the measurements will not be present in $\mathbf{S}^H$. This increases the likelihood that the target will be missed by the radar sensor as the signal element corresponding to the true target state will exist in the null space of the sensing matrix.

B.2.2 NON-ADAPTIVE CSP (NACSP)

However, prior information on the target state is ignored when using NACSP. The benefit of NACSP lies in the universality of its application and comes to the expense of a lower SNR. In NACSP, the matrix $\mathbf{\Phi}$ is size $C \times M$ matrix with $C \ll M$ chosen to be populated with random entries and orthonormalized rows having the restricted isometry property (RIP) [82]. As shown in [82],

$$\mathbf{g}_{\bar{l}}^H \mathbf{g}_l \approx \begin{cases} \frac{C}{M} \tilde{\boldsymbol{\delta}}_{\bar{l}}^H \tilde{\boldsymbol{\delta}}_l, & \bar{l} \neq l \\ \frac{C}{M}, & \bar{l} = l \end{cases} \tag{B.5}$$

and

$$\mathbf{g}_l^H \mathbf{\Phi} \mathbf{\Phi}^H \mathbf{g}_l \approx \frac{C}{M}. \tag{B.6}$$

The SNR in (3.11) using (B.5) and (B.6) in the NACSP then becomes

$$\eta_{l,l,u,k} = \frac{C}{M} \frac{\sigma_\gamma^2}{\sigma_v^2}. \tag{B.7}$$

B.2.3 NYQUIST SENSING AND PROCESSING (NSP)

The best SNR performance is achieved by NSP, at the expense of increasing the dimensionality of the received signal to its Nyquist rate dimensionality M. In NSP can be seen to be a special case of ACSP by setting $C = M$. The sensing and processing matrix becomes $\mathbf{\Phi} = \mathbf{S}^H$ and then

$$\mathbf{g}_{\bar{l}}^H \mathbf{g}_l = \begin{cases} \mathbf{s}_{\bar{l}}^H \mathbf{s}_l, & \bar{l} \neq l \\ \xi_s, & \bar{l} = l. \end{cases} \tag{B.8}$$

The SNR in (3.11) using (B.8) becomes in NSP equal to the input SNR given by (3.12)

$$\eta_{l,l,u,k} = \frac{\sigma_\gamma^2 \xi_s}{\sigma_v^2}. \tag{B.9}$$

It can be observed from (B.7) and (B.9) that the SNR in the case of the NACSP is scaled by $\frac{C}{M}$ compared to NSP which comes at a cost of sampling and processing at higher rate and dimensionality. On the other hand, from (B.4) and (B.9) it can be seen that in ASCSP, for example, the SNR is scaled by only $\frac{C}{\tilde{L}_{\mathrm{b},u,k}}$ compared to NSP where $\tilde{L}_{\mathrm{b},u,k} < M$ demonstrating the benefit provided by ASCSP.

Bibliography

[1] S. Haykin, Cognitive radar: A way of the future, *IEEE Signal Processing Magazine*, vol. 23, no. 1, pp. 30–40, 2006. DOI: 10.1109/msp.2006.1593335 1, 2

[2] I. Kyriakides, Target tracking using adaptive compressive sensing and processing, *Signal Processing*, vol. 127, 2016. DOI: 10.1016/j.sigpro.2016.02.019 1, 3, 10, 27, 28, 29, 33, 43, 44

[3] S. Haykin, Cognitive dynamic systems [Point of View], *Proc. IEEE*, vol. 94, no. 11, pp. 1910–1911, 2006. 1

[4] S. Haykin, Cognitive dynamic systems: Radar, control, and radio [Point of View], *Proc. of the IEEE*, vol. 100, no. 7, pp. 2095–2103, July 2012. DOI: 10.1109/jproc.2012.2193709

[5] S. Haykin, Cognitive networks: Radar, radio, and control for new generation of engineered complex networks, *IEEE National Radar Conference Proceedings*, 2013. DOI: 10.1109/radar.2013.6586147 1

[6] S. Haykin, Cognitive radio: Brain-empowered wireless communications, *IEEE Journal on Selected Areas in Communications*, vol. 23, no. 2, pp. 201–220, February 2005. DOI: 10.1109/jsac.2004.839380

[7] J. R. Guerci, Cognitive radar: A knowledge-aided fully adaptive approach, *IEEE Radar Conference*, pp. 1365–1370, 2010. DOI: 10.1109/radar.2010.5494403

[8] M. Fatemi and S. Haykin, Cognitive control: Theory and application, *IEEE Access*, vol. 2, pp. 698–710, 2014. DOI: 10.1109/access.2014.2332333 1

[9] S. P. Sira, A. Papandreou-Suppappola, and D. Morrell, *Advances in Waveform-Agile Sensing for Tracking*, Morgan & Claypool Publishers, 2009. DOI: 10.2200/S00168ED1V01Y200812ASE002 1, 3

[10] D. Morrell and Y. Xue, Adaptive foveal sensor for target tracking, *Asilomar Conference on Signals, Systems and Computers*, pp. 848–852, 2002. DOI: 10.1109/acssc.2002.1197298 1, 2, 3, 25

[11] I. Kyriakides, D. Morrell, and A. Papandreou-Suppappola, Using a configurable integrated sensing and processing imager to track multiple targets, *IEEE Aerospace Conference Proceedings*, 2008. DOI: 10.1109/aero.2008.4526446 1

[12] A. Saksena and I-J. Wang, Dynamic ping optimization for surveillance in multistatic sonar buoy networks with energy constraints, *47th IEEE Conference on Decision and Control*, pp. 1109–1114, 2008. DOI: 10.1109/cdc.2008.4739329 1

[13] P. Chavali and A. Nehorai, Scheduling and power allocation in a cognitive radar network for multiple-target tracking, *IEEE Transactions on Signal Processing*, vol. 60, no. 2, pp. 715–729, 2012. DOI: 10.1109/tsp.2011.2174989 1

[14] A. O. Hero and C. M. Kreucher, Network sensor management for tracking and localization, *10th International Conference on Information Fusion*, pp. 1–8, IEEE, July 2007. DOI: 10.1109/icif.2007.4408181 1

[15] E. Fishler, A. Haimovich, R. S. Blum, L. J. Cimini, D. Chizhik, and R. A. Valenzuela, Spatial diversity in radars—models and detection performance, *IEEE Transactions on Signal Processing*, vol. 54, no. 3, pp. 823–838, March 2006. DOI: 10.1109/tsp.2005.862813

[16] I. Bekkerman and J. Tabrikian, Target detection and localization using MIMO radars and sonars, *IEEE Transactions on Signal Processing*, vol. 54, no. 10, pp. 3873–3883, October 2006. DOI: 10.1109/tsp.2006.879267

[17] R. Niu, R. S. Blum, P. K. Varshney, and A. L. Drozd, Target localization and tracking in noncoherent multiple-input multiple-output radar systems, *IEEE Transactions on Aerospace and Electronic Systems*, vol. 48, no. 2, pp. 1466–1489, 2012. DOI: 10.1109/taes.2012.6178073 1

[18] K. L. Bell, C. J. Baker, G. E. Smith, J. T. Johnson, and M. Rangaswamy, Cognitive radar framework for target detection and tracking, *IEEE Journal of Selected Topics in Signal Processing*, vol. 9, no. 8, pp. 1427–1439, December 2015. DOI: 10.1109/jstsp.2015.2465304 2

[19] K. L. Bell, C. J. Baker, G. E. Smith, J. T. Johnson, and M. Rangaswamy, Fully adaptive radar for target tracking part I: Single target tracking, *IEEE Radar Conference*, pp. 0303–0308, May 2014. DOI: 10.1109/radar.2014.6875604

[20] K. L. Bell, C. J. Baker, G. E. Smith, J. T. Johnson, and M. Rangaswamy, Fully adaptive radar for target tracking part II: Target detection and track initiation, *IEEE Radar Conference*, pp. 0309–0314, May 2014. DOI: 10.1109/radar.2014.6875605 2

[21] S. Haykin, A. Zia, I. Arasaratnam, and Y. Xue, Cognitive tracking radar, *IEEE Radar Conference*, pp. 1467–1470, 2010. DOI: 10.1109/radar.2010.5494383 2

[22] S. Haykin, Y. Xue, and P. Setoodeh, Cognitive radar: Step toward bridging the gap between neuroscience and engineering, *Proc. of the IEEE*, vol. 100, no. 11, pp. 3102–3130, November 2012. DOI: 10.1109/jproc.2012.2203089 2

[23] S. Haykin, Y. Xue, and T. N. Davidson, Optimal waveform design for cognitive radar, *42nd Asilomar Conference on Signals, Systems and Computers*, pp. 3–7, IEEE, October 2008. DOI: 10.1109/acssc.2008.5074349 2, 3, 27

[24] D. J. Kershaw and R. J. Evans, Waveform selective probabilistic data association, *IEEE Transactions on Aerospace and Electronic Systems*, vol. 33, no. 4, pp. 1180–1188, October 1997. DOI: 10.1109/7.625110

[25] S. P. Sira, Y. Li, A. Papandreou-Suppappola, D. Morrell, D. Cochran, and M. Rangaswamy, Waveform-agile sensing for tracking, *IEEE Signal Processing Magazine*, vol. 26, no. 1, pp. 53–64, January 2009. DOI: 10.1109/msp.2008.930418 3, 18

[26] S. P. Sira, A. Papandreou-Suppappola, and D. Morrell, Dynamic configuration of time-varying waveforms for agile sensing and tracking in clutter, *IEEE Transactions on Signal Processing*, vol. 55, no. 7, pp. 3207–3217, July 2007. DOI: 10.1109/tsp.2007.894418 2, 27

[27] C. Kreucher and K. Carter, An information theoretic approach to processing management, *IEEE International Conference on Acoustics, Speech and Signal Processing*, pp. 1869–1872, March 2008. DOI: 10.1109/icassp.2008.4517998 2, 3

[28] K. L. Bell, J. T. Johnson, G. E. Smith, C. J. Baker, and M. Rangaswamy, Cognitive radar for target tracking using a software defined radar system, *IEEE Radar Conference (RadarCon)*, pp. 1394–1399, May 2015. DOI: 10.1109/radar.2015.7131213 2

[29] H. Sun-mog, R. J. Evans, and S. Han-seop, Optimization of waveform and detection threshold for range and range-rate tracking in clutter, *IEEE Transactions on Aerospace and Electronic Systems*, vol. 41, no. 1, pp. 17–33, January 2005. DOI: 10.1109/taes.2005.1413743 3

[30] B. F. La Scala, W. Moran, and R. J. Evans, Optimal adaptive waveform selection for target detection, *Proc. of the International Conference on Radar (IEEE Cat. no. 03EX695)*, pp. 492–496, 2003. DOI: 10.1109/radar.2003.1278791

[31] U. Gunturkun, Toward the development of radar scene analyzer for cognitive radar, *IEEE Journal of Oceanic Engineering*, vol. 35, no. 2, pp. 303–313, April 2010. DOI: 10.1109/joe.2010.2043378 3

[32] M. Hurtado, T. Zhao, and A. Nehorai, Adaptive polarized waveform design for target tracking based on sequential Bayesian inference, *IEEE Transactions on Signal Processing*, vol. 56, no. 3, pp. 1120–1133, 2008. DOI: 10.1109/tsp.2007.909044 3

[33] P. Chavali and A. Nehorai, Cognitive radar for target tracking in multipath scenarios, *International Waveform Diversity and Design Conference*, pp. 000110–000114, IEEE, August 2010. DOI: 10.1109/wdd.2010.5592379 3

[34] F. Gustafsson, Particle filter theory and practice with positioning applications, *IEEE Aerospace and Electronic Systems Magazine*, vol. 25, no. 7, pp. 53–82, July 2010. DOI: 10.1109/maes.2010.5546308 3, 4, 13, 18

[35] I. Kyriakides, D. Morrell, and A. Papandreou-Suppappola, Adaptive highly localized waveform design for multiple target tracking, *EURASIP Journal on Advances in Signal Processing*, vol. 2012, no. 1, 2012. DOI: 10.1186/1687-6180-2012-180 3

[36] R. A. Romero and N. A. Goodman, Cognitive radar network: Cooperative adaptive beamsteering for integrated search-and-track application, *IEEE Transactions on Aerospace and Electronic Systems*, vol. 49, no. 2, pp. 915–931, April 2013. DOI: 10.1109/taes.2013.6494389 3

[37] W. Huleihel, J. Tabrikian, and R. Shavit, Optimal adaptive waveform design for cognitive MIMO radar, *IEEE Transactions on Signal Processing*, vol. 61, no. 20, 2013. DOI: 10.1109/tsp.2013.2269045

[38] N. Sharaga, J. Tabrikian, and H. Messer, Optimal cognitive beamforming for target tracking in MIMO radar/sonar, *IEEE Journal of Selected Topics in Signal Processing*, vol. 9, no. 8, pp. 1440–1450, December 2015. DOI: 10.1109/jstsp.2015.2467354 3

[39] S. Xia, R. Sridhar, P. Scott, and C. Bandera, An all CMOS foveal image sensor chip, *Proc. 11th Annual IEEE International ASIC Conference (Cat. no. 98TH8372)*, pp. 409–413, IEEE. DOI: 10.1109/asic.1998.723085 3, 25

[40] J. Batista, P. Peixoto, and H. Araujo, Real-time active visual surveillance by integrating peripheral motion detection with foveated tracking, *Proc. IEEE Workshop on Visual Surveillance*, pp. 18–25, IEEE Computer Society, 1998. DOI: 10.1109/wvs.1998.646016

[41] Y. Xue and D. Morrell, Target tracking and data fusion using multiple adaptive Foveal sensors, *6th International Conference of Information Fusion, Proc. of the*, pp. 326–333, IEEE, 2003. DOI: 10.1109/icif.2003.177464 25

[42] G. Spell and D. Cochran, Algorithms for tracking with a foveal sensor, *Asilomar Conference on Signals, Systems and Computers*, pp. 1563–1565, 2015. DOI: 10.1109/acssc.2015.7421409 3

[43] D. Cochran and R. Martin, Nonlinear filtering models of attentive vision, *IEEE International Symposium on Circuits and Systems. Circuits and Systems Connecting the World. ISCAS*, vol. Supplement, pp. 26–29, 1996. DOI: 10.1109/iscas.1996.598468 3

[44] L. Lit, D. Cochrant, and R. Martint, Target tracking with an attentive foveal sensor, *34th Asilomar Conference on Signals, Systems and Computers*, pp. 182–185, 2000. DOI: 10.1109/acssc.2000.910940 3, 25

[45] C. M. Kreucher, A. O. Hero, K. D. Kastella, and M. R. Morelande, An information-based approach to sensor management in large dynamic networks, *Proc. of the IEEE*, vol. 95, no. 5, pp. 978–999, May 2007. DOI: 10.1109/jproc.2007.893247 3

[46] J. Zhang, D. Zhu, and G. Zhang, Adaptive compressed sensing radar oriented toward cognitive detection in dynamic sparse target scene, *IEEE Transactions on Signal Processing*, vol. 60, no. 4, pp. 1718–1729, 2012. DOI: 10.1109/tsp.2012.2183127 3, 40

[47] D. Lahat, T. Adali, and C. Jutten, Multimodal data fusion: An overview of methods, challenges, and prospects, *Proc. of the IEEE*, vol. 103, no. 9, pp. 1449–1477, September 2015. DOI: 10.1109/jproc.2015.2460697 3

[48] M. S. Arulampalam, S. Maskell, N. Gordon, and T. Clapp, A tutorial on particle filters for online nonlinear/non-Gaussian Bayesian tracking, *IEEE Transactions on Signal Processing*, vol. 50, no. 2, pp. 174–188, 2002. DOI: 10.1109/78.978374 3, 4, 9, 15, 21, 22, 33, 34, 41

[49] M. Orton and W. Fitzgerald, A Bayesian approach to tracking multiple targets using sensor arrays and particle filters, *IEEE Transactions on Signal Processing*, vol. 50, no. 2, pp. 216–223, 2002. DOI: 10.1109/78.978377 4

[50] C. Kreucher, K. Kastella, and A. O. Hero, Multitarget tracking using the joint multitarget probability density, *IEEE Transactions on Aerospace and Electronic Systems*, vol. 41, no. 4, pp. 1396–1414, october 2005. DOI: 10.1109/taes.2005.1561892 4

[51] I. Kyriakides, D. Morrell, and A. Papandreou-Suppappola, Sequential Monte Carlo methods for tracking multiple targets with deterministic and stochastic constraints, *IEEE Transactions on Signal Processing*, vol. 56, no. 3, 2008. DOI: 10.1109/tsp.2007.908931 3, 4

[52] W. Burgard, F. Dellaert, D. Fox, and S. Thrun, Particle filters for mobile robot localization, *Sequential Monte Carlo Methods in Practice*, de Freitas N. Gordon Doucet, A., Ed., Springer, NY, 2001. DOI: 10.1007/978-1-4757-3437-9_19 4

[53] P. Torma and C. Szepesvári, Enhancing particle filters using local likelihood sampling, *ECCV2004 Lecture Notes in Computer Science*, pp. 16–28, 2004. DOI: 10.1007/978-3-540-24670-1_2 4

[54] I. Kyriakides, Adaptive compressive sensing and processing for radar tracking, *IEEE International Conference on Acoustics, Speech and Signal Processing*, 2011. DOI: 10.1109/icassp.2011.5947201 4, 28, 40, 43

[55] S. Singh, B.-N. Vo, and A. Doucet, Sequential Monte Carlo methods for multitarget filtering with random finite sets, *IEEE Transactions on Aerospace and Electronic Systems*, vol. 41, no. 4, pp. 1224–1245, 2005. DOI: 10.1109/taes.2005.1561884 5

[56] R. Mahler, *Advances in Statistical Multisource-Multitarget Information Fusion*, Artech House, Boston, 2014.

[57] R. Mahler, An introduction to multisource-multitarget statistics and applications, *Technical Report*, Lockheed Martin Technical Monograph, 2000.

[58] T. Zajic and R. Mahler, A particle-systems implementation of the PHD multi-target tracking filter, *Signal Processing, Sensor Fusion and Target Recognition XII*, SPIE, Ed., pp. 291–299, 2003. DOI: 10.1117/12.488533 5

[59] M. Clark, S. Maskell, R. Vinter, and M. Yaqoob, A comparison of the particle and shifted Rayleigh filters in their application to a multi-sensor bearings-only problem, *IEEE Aerospace Conference Proceedings*, 2005. DOI: 10.1109/aero.2005.1559505 5

[60] I. Kyriakides, Ambiguity function surface when using prior information in compressive sensing and processing, *3rd International Workshop on Compressed Sensing Theory and its Applications to Radar, Sonar, and Remote Sensing, CoSeRa*, 2015. DOI: 10.1109/cosera.2015.7330268 11, 28, 40, 43

[61] C. Kreucher, M. Morelande, K. Kastella, and A. O. Hero, Particle filtering for multi-target detection and tracking, *IEEE Aerospace Conference*, pp. 2101–2116, 2005. DOI: 10.1109/aero.2005.1559502 12

[62] D. Morrell, I. Kyriakides, and A. Papandreou-Suppappola, On the validity of the measurement independence assumption when using CAZAC and LFM waveforms in radar tracking with a particle filter, *Technical Report*, Dept. of Electrical Engineering, Arizona State University, 2007.

[63] D. Morrell, I. Kyriakides, and A. Papandreou-Suppappola, On the validity of the measurement independence approximation when using single and MCPC waveforms based on Björck CAZAC sequences in multiple target radar tracking with a particle filter, *Technical Report*, Dept. of Electrical Engineering, Arizona State University, 2008. 12

[64] I. Kyriakides, Target tracking using cognitive radar and foveal nodes, *15th IEEE International Conference on Distributed Computing in Sensor Systems and Workshops*, 2019. DOI: 10.1109/dcoss.2019.00124 25, 34

[65] M. Bolduc and M. D. Levine, A real-time foveated sensor with overlapping receptive fields, *Journal of Real-Time Imaging*, vol. 3, no. 3, pp. 195–212, 1997. DOI: 10.1006/rtim.1996.0056 25

[66] G. Sandini, P. Questa, D. Scheffer, B. Diericks, and A. Mannucci, A retina-like CMOS sensor and its applications, *IEEE Sensor Array and Multichannel Signal Processing Workshop*, pp. 514–519, 2000. DOI: 10.1109/sam.2000.878062 25

[67] H. Strasburger, I. Rentschler, and M. Juttner, Peripheral vision and pattern recognition: A review, *Journal of Vision*, vol. 11, no. 5, pp. 13–13, 2011. DOI: 10.1167/11.5.13 26

[68] D. B. Phillips, Ming-Jie Sun, J. M. Taylor, M. P. Edgar, S. M. Barnett, G. G. Gibson, and M. J. Padgett, Adaptive foveated single-pixel imaging with dynamic super-sampling, *Science Advances*, vol. 3, no. 4, 2017. DOI: 10.1126/sciadv.1601782 26

[69] R. P. de Figueiredo, A. Bernardino, J. Santos-Victor, and H. Araújo, On the advantages of foveal mechanisms for active stereo systems in visual search tasks, *Autonomous Robots*, pp. 1–18, 2017. DOI: 10.1007/s10514-017-9617-1 26

[70] M. I. Skolnik, *Introduction to Radar Systems*, McGraw-Hill, 2001. 27, 30

[71] J. Romberg, E. J. Candes, and T. Tao, Robust uncertainty principles: Exact signal reconstruction from highly incomplete frequency information, *IEEE Transactions on Information Theory*, vol. 52, no. 2, pp. 489–509, 2006. DOI: 10.1109/tit.2005.862083 28, 39

[72] E. Candes and M. B. Wakin, An introduction to compressive sampling, *IEEE Signal Processing Magazine*, vol. 25, no. 2, pp. 21–30, 2008. DOI: 10.1109/msp.2007.914731 29

[73] D. Donoho, Compressed sensing, *IEEE Transactions on Information Theory*, no. 4, pp. 1289–1306, 2006. DOI: 10.1109/tit.2006.871582

[74] R. G. Baraniuk, Compressive sensing, *IEEE Signal Processing Magazine*, vol. 24, pp. 118–121, 2007. DOI: 10.1109/ciss.2008.4558479 28, 39

[75] R. Baraniuk and P. Steeghs, Compressive radar imaging, *Proc. IEEE Radar Conference*, pp. 128–133, 2007. DOI: 10.1109/radar.2007.374203 40

[76] M. F. Duarte, D. Baron, J. A. Tropp, M. B. Wakin, and R. G. Baraniuk, Random filters for compressive sampling and reconstruction, *IEEE International Conference on Acoustics, Speech, and Signal Processing*, vol. 3, 2006. DOI: 10.1109/icassp.2006.1660793 28

[77] A. C. Gilbert and J. A. Tropp, Signal recovery from random measurements via orthogonal matching pursuit, *IEEE Transactions on Information Theory*, vol. 53, pp. 4655–4666, 2007. DOI: 10.1109/tit.2007.909108

[78] J. A. Tropp and S. J. Wright, Computational methods for sparse solution of linear inverse problems, *IEEE Proc.*, vol. 98, no. 6, pp. 948–958, 2010. DOI: 10.21236/ada633835 40

[79] E. van den Berg and M. P. Friedlander, SPGL1: A solver for large-scale sparse reconstruction, `http://www.cs.ubc.ca/~mpf/spgl1`, 2007. 40

[80] E. van den Berg and M. P. Friedlander, Probing the Pareto frontier for basis pursuit solutions, *SIAM Journal on Scientific Computing*, vol. 31, no. 2, pp. 890–912, 2008. DOI: 10.1137/080714488

[81] L. Carin, Ji Shihao, and Xue Ya, Bayesian compressive sensing, *IEEE Transactions on Signal Processing*, vol. 56, pp. 2346–2356, 2008. DOI: /10.1109/tsp.2007.914345 28

[82] M. B. Wakin, M. A. Davenport, P. T. Boufounos, and R. G. Baraniuk, Signal processing with compressive measurements, *IEEE Journal of Selected Topics in Signal Processing*, vol. 4, no. 2, pp. 445–460, 2010. DOI: 10.1109/jstsp.2009.2039178 28, 43, 44, 45

[83] I. Kyriakides, A configurable compressive acquisition matrix for radar tracking using particle filtering, *9th European Radar Conference, EuRAD*, 2012. 28, 43

[84] R. M. Castro, J. D. Haupt, R. G. Baraniuk, and R. D. Nowak, Compressive distilled sensing: Sparse recovery using adaptivity in compressive measurements, *Asilomar Conference Signals, Systems and Computers*, pp. 1551–1555, 2009. DOI: 10.1109/acssc.2009.5470138 28, 40

[85] R. Nowak, J. Haupt, and R. M. Castro, Distilled sensing: Adaptive sampling for sparse detection and estimation, *IEEE Transactions on Information Theory*, vol. 57, no. 9, pp. 6222–6235, 2011. DOI: 10.1109/tit.2011.2162269 28, 40

[86] G. Björck, Functions of modulus one on Zn whose Fourier transforms have constant modulus, and cyclic n-roots, *Proc. of NATO Advanced Study Institute on Recent Advances in Fourier Analysis and its Applications*, pp. 131–140, J. S. Byrnes and J. L. Byrnes, 1990. DOI: 10.1007/978-94-009-0665-5_10 29, 36

[87] I. Kyriakides, I. Konstantinidis, D. Morrell, J. J. Benedetto, and A. Papandreou-Suppappola, Target tracking using particle filtering and CAZAC sequences, *International Waveform Diversity and Design Conference*, 2007. DOI: 10.1109/wddc.2007.4339445 29, 36

[88] H. Braun, P. Turaga, A. Spanias, and C. Tepedelenlioglu, Methods, Apparatuses, and Systems for Reconstruction-Free Image Recognition from Compressive Sensors, U.S. 10,387,751, 2019. 40

[89] H. Braun, P. Turaga, A. Spanias, S. Katoch, S. Jayasuriya, and C. Tepedelenlioglu, *Reconstruction-Free Compressive Vision for Surveillance Applications,* Morgan & Claypool Publishers, 2019. 40

[90] M. Stanley and J.M. Lee, *Sensors for IoT Applications,* Andress Spanias, Ed., Morgan & Claypool Publishers, 2018. 40

[91] S. Zhang, C. Tepedelenlioglu, A. Spanias, and M. Banavar, *Distributed Network Structure Estimation using Consensus Methods,* William Tranter, Ed., Morgan & Claypool Publishers, 2018. 40

Author's Biography

IOANNIS KYRIAKIDES

Ioannis Kyriakides received his B.S. in Electrical Engineering in 2003 from Texas A&M University. He received his M.S. and Ph.D. in 2005 and 2008, respectively, from Arizona State University. He held a research associate position throughout his graduate studies, funded by the Integrated Sensing and Processing program and the Multidisciplinary University Research Initiative of the U.S. Department of Defense. In the final year of his Ph.D. work, he received the University Graduate Fellowship of the Arizona State University. His research interests include Bayesian target tracking, sequential Monte Carlo methods, and cognitive fusion for heterogeneous data fusion and heterogeneous sensing node configuration. His research work includes localization of multiple RF sources, tracking surface vehicles using passive acoustic sensing, tracking multiple targets with constraints in motion, and tracking multiple targets using heterogeneous cognitive sensors. Dr. Kyriakides is currently an Associate Professor at the Engineering Department at the University of Nicosia. He participates in the Department of Engineering Quality Assurance Committee and the University of Nicosia Environment, Health, and Safety Committee. He is a member of CapTech Information of the European Defence Agency. He has been the coordinator of a Research and Innovation Foundation project "Automated Sonar-Thermal Monitoring for Sea Border Security" with budget 180 K Euro. He is currently the coordinator of the Research and Innovation Foundation Project "MARI-Sense: Maritime Cognitive Decision Support System" with budget 1 M Euro (`www.marisense.eu`).